Contributions to Statistics

Fulvia Mecatti • Pier Luigi Conti •
Maria Giovanna Ranalli

Editors

Contributions to Sampling Statistics

 Springer

Editors
Fulvia Mecatti
Dept. of Sociology and Social Research
University of Milano-Bicocca
Milano
Italy

Pier Luigi Conti
Dept. of Statistical Sciences
Sapienza - University of Rome
Roma
Italy

Maria Giovanna Ranalli
Dept. Economics, Finance and Statistics
University of Perugia
Perugia
Italy

ISSN 1431-1968
ISBN 978-3-319-35857-4   ISBN 978-3-319-05320-2 (eBook)
DOI 10.1007/978-3-319-05320-2
Springer Cham Heidelberg New York Dordrecht London

Preface

This book comprises a selection of the papers presented at the Third Italian Conference on Survey Methodology, *ITACOSM 2013*, held in Milan in June 2013 (http://www.statistica.unimib.it/itacosm13/).

ITACOSM is the bi-annual international meeting of S^2G, the Survey Sampling Group of the Italian Statistical Society. It is intended as a forum for scientific discussion on developments in the theory and application of survey sampling methodologies for the human and natural sciences.

The volume gathers 14 contributions that address a wide range of sampling methods and techniques central to international statistical research. Three invited papers by keynote speakers at the conference open the book, offering state-of-the-art reviews of some hot topics in sampling theory: calibration for nonresponse treatment (P. S. Kott), quantile regression methods for small area estimation (M. Pratesi) and estimation from multiple frame surveys (A.C. Singh). Eleven research papers follow, carefully selected from both invited and contributed sessions of the conference and listed in alphabetical order by author. The contents of these papers range across mainstream topics in both sampling theory and applications. Theoretical contributions consider inference from complex samples, randomized response techniques, bootstrap methods, weighting, modeling, imputation and effective use of auxiliary information; applications cover a varied collection of subjects in official household surveys, Bayesian networks for measurement error, auditing, mixed-mode surveys, business and economic surveys, agricultural data and geostatistics.

The book is intended to serve as a "hot topics reference" for survey sampling researchers, professionals and practitioners in a wide array of application fields. It brings to a close a project that, had I known in advance, I would probably not have committed myself to! As president of S^2G, I first took charge of the organization of ITACOSM 2013 at University of Milan-Bicocca in April 2011, with a tiny budget and no experience. Afterwards, it is something of which I am enormously glad, proud and grateful. My sincere gratitude is due to many people: to all who supported me in the organization of ITACOSM 2013, particularly Federica Nicolussi and Paola Chiodini; to all the authors, speakers and attendants; to my co-editors Maria

Giovanna Ranalli and Pier Luigi Conti, who did a great job in the difficult task of selecting from among the many submitted manuscripts; and to the anonymous referees.

This book is published thanks to a grant by Marcello Fontanesi, past-Chancellor of the University of Milan-Bicocca.

Milan, Italy   Fulvia Mecatti
20 January 2014

Contents

Calibration Weighting When Model and Calibration Variables Can Differ

Phillip S. Kott

Abstract Calibration weighting is an easy-to-implement yet powerful tool for reducing the standard errors of many population estimates derived from a sample survey by forcing the weighted sums of certain "calibration" variables to equal their known (or better-estimated) population totals. Although originally developed to reduce standard errors, calibration weighting can also be used to reduce or remove selection biases resulting from unit nonresponse. To this end, nonrespondents are usually assumed to be "missing at random," that is, the response mechanism is assumed to be a function of calibration variables with either known values in the entire sample or known population totals. It is possible, however, to use calibration-weighting to compensate for unit nonresponse when response is a function of model variables that need not be calibration variables; in fact, some model variables can have values known only for respondents. We will explore some recent findings connected with this methodology.

Keywords General exponential model • Generalized raking • Nonresponse bias • Response model • Variance estimation

1 Introduction

Survey sampling is a tool used primarily for estimating the parameters of a finite population based on a randomly drawn sample of its members. Probability samples, in which members have differing probabilities of sample selection, come with sampling weights, which are often the inverses of the individual member selection probabilities. As long as each population member has a positive selection

P.S. Kott (✉)
RTI International, 6110 Executive Blvd., Suite 902, Rockville, MD 20852, USA
e-mail: pkott@rti.org

F. Mecatti et al. (eds.), *Contributions to Sampling Statistics*, Contributions to Statistics,
DOI 10.1007/978-3-319-05320-2__1,

probability (and therefore a finite sampling weight), it is a simple matter to estimate a population total in an unbiased fashion. Population means and ratios of population totals can also usually be estimated in a manner that is nearly unbiased in some sense. Calibration weighting (Deville and Särndal 1992) is a tool for modestly adjusting sampling weights in such a way that the weighted sums of certain "calibration" variables equal their known (or better-estimated) population totals. As a consequence of these *calibration equations* holding, the parameter estimates for variables without known population totals remain nearly unbiased while their standard errors are often reduced.

Although originally developed to reduce standard errors, calibration weighting has also been effectively used to remove selection biases resulting from unit nonresponse (e.g., Folsom 1991; Fuller et al. 1994; Lundström and Särndal 1999; Folsom and Singh 2000; Kott 2006). To this end, whether a population member that has been selected for the sample responds to a survey is treated as an additional phase of Poisson random sampling with positive, but otherwise unknown, selection probabilities. Calibration weighting can then be used to estimate these Poisson selection probabilities implicitly and produce parameter estimates that are nearly unbiased under *quasi-probability-sampling theory*. An important *caveat* is that although the sample-selection mechanism is fully under the control of the statistician, the response-selection mechanism is unknown and only assumed to have a particular form, an assumption which can fail allowing the resulting estimators to be biased.

In most applications of calibration weighting for nonresponse adjustment, nonrespondents are assumed to be "missing at random" (Rubin 1976), which means in our context that the response mechanism is assumed to be a function of variables having known totals for either the entire population or the sampling-weighted sample before nonresponse.

Deville (2000) pointed out that, contrary to standard practice, weighting can be used to remove the biases from unit nonresponse that is wholly or in part a function of variables with values known only for respondents. Chang and Kott (2008) not only added rigor to Deville's observation but extended the notion of calibration weighting by allowing the number of explanatory variables in the assumed response model to be less than the number of calibration variables.

When there is more calibration than response-model variables, the weighted sums of all the calibration variables cannot be forced to equal their population or full-sample targets. Instead, the calibration-weighting method in this context only forces these sums to come close to their targets.

We will explore some theoretical results concerning calibration weighting when the model and calibration variables can differ. Concepts will be stressed rather than rigorous proofs. For that, there are references, such as Fuller (2009). The term "under mild conditions" will be used repeatedly but not formally defined. In the final section, we will briefly discuss what some of those conditions are and why they might fail.

2   Linear Calibration Weighting in the Purest Form

In the absence of nonresponse or frame errors, calibration weighting is a weight-adjustment method that creates a set of weights, $\{w_k\}$, for elements in the sample S that:

1. Are asymptotically close to the original probability-sampling weights: $d_k = 1/\pi_k$ (i.e., as the sample size grows arbitrarily large, w_k converges to d_k).
2. Satisfy a finite set of calibration equations (one for each component of $\mathbf{z}_k$):

$$\sum_{k \in S} w_k \mathbf{z}_k = \sum_{k \in U} \mathbf{z}_k,$$

where U denotes the population.

When estimating $T = \sum_U y_k$ with $t_{\text{CAL}} = \sum_S w_k y_k$, *calibration weighting* will be nearly unbiased under probability-sampling theory while tending to reduce mean squared error relative to the simple expansion estimator, $t_{\text{EXP}} = \sum_S d_k y_k$, when y_k is correlated with components of $\mathbf{z}_k$.

Although we focus here on estimating a single y-total here, most surveys are devised to estimate more than one population total (or population mean). As a result, it is possible that calibration weighting will reduce the mean squared errors of some estimated totals but not others. Usually, the vector $\mathbf{z}_k$ is chosen so that few estimated totals of interest to the survey designers have increased mean squared errors.

The simplest way to compute calibration weights is linearly with the following formula:

$$w_k = d_k \left[1 + \left(\sum_{j \in U} \mathbf{z}_j - \sum_{j \in S} d_j \mathbf{z}_j \right)^T \left(\sum_{j \in S} d_j \mathbf{z}_j \mathbf{z}_j^T \right)^{-1} \mathbf{z}_k \right]$$

$$= d_k \left[1 + \mathbf{g}^T \mathbf{z}_k \right],$$

where

$$\mathbf{g} = \left(\sum_{j \in S} d_j \mathbf{z}_j \mathbf{z}_j^T \right)^{-1} \left(\sum_{j \in U} \mathbf{z}_j - \sum_{j \in S} d_j \mathbf{z}_j \right).$$

Observe that as the sample size grows arbitrarily large, $(\sum_U \mathbf{z}_j - \sum_S d_j \mathbf{z}_j)/N$, where N is the population size, converges to $\mathbf{0}$ under mild conditions and thus so does $\mathbf{g}$ and the scalar $\mathbf{g}^T \mathbf{z}_k$. This means that under those mild conditions, the relative difference between the calibration estimator and the expansion estimator itself converges to $\mathbf{0}$ asymptotically. Since t_{EXP} is unbiased under probability sampling theory, t_{CAL} is *nearly* unbiased under probability-sampling theory: its relative bias tends to zero asymptotically.

The linear calibration weighting scheme described above implies the *generalized regression* (GREG) estimator,

$$t_{\text{GREG}} = \sum_{k \in S} w_k y_k = \sum_{k \in S} d_k y_k + \left(\sum_{j \in U} \mathbf{z}_j - \sum_{j \in S} d_j \mathbf{z}_j \right)^T \mathbf{b}_S, \tag{1}$$

where $\mathbf{b}_S = (\sum_S d_j \mathbf{z}_j \mathbf{z}_j^T)^{-1} \sum_S d_k \mathbf{z}_k y_k$ is a probability-sampling-weighted regression coefficient. Kott (2009) provides a good review of the properties of a calibration-weighted estimator and its relationship to the GREG.

3   The GREG and Unit Nonresponse

Following Fuller et al. (1994), the GREG can be used to handle unit nonresponse in the following manner: simply replace the sample S by the respondent sample R in defining the GREG with

$$t_{\text{GREG}} = \sum_{k \in R} w_k y_k = \sum_{k \in R} d_k \left(1 + \mathbf{g}^T \mathbf{z}_k \right) y_k,$$

where now

$$\mathbf{g} = \left(\left[(1 - \theta) \sum_{j \in U} \mathbf{z}_j + \theta \sum_{j \in S} d_j \mathbf{z}_j \right] - \sum_{j \in R} d_j \mathbf{z}_j \right)^T \left(\sum_{j \in R} d_j \mathbf{z}_j \mathbf{z}_j^T \right)^{-1},$$

and θ is either 0 or 1. When $\theta = 0$, the respondent sample is said to be *calibrated to the population*. When $\theta = 1$, the respondent sample is said to be *calibrated to the sample*.

We will discuss some theoretical reasons for preferring one over the other later. In practice, the choice may depend on which total is available: $\Sigma_U \mathbf{z}_j$ or $\Sigma_S d_j \mathbf{z}_j$. In fact, some component totals may only be available at the sample level while others are only available at the population level. It would be a simple, but cumbersome, matter to write $\mathbf{z}_j^T$ as $(\mathbf{z}_{1j}^T \ \mathbf{z}_{2j}^T)$ where $\Sigma_U \mathbf{z}_{1j}$ and $\Sigma_S d_j \mathbf{z}_{2j}$ are known, and replace $\mathbf{g}$ with

$$\mathbf{g} = \left(\begin{array}{c} \sum_{j \in U} \mathbf{z}_{1j} - \sum_{j \in R} d_j \mathbf{z}_{1j} \\ \sum_{j \in S} d_j \mathbf{z}_{2j} - \sum_{j \in R} d_j \mathbf{z}_{2j} \end{array} \right)^T \left(\sum_{j \in R} d_j \mathbf{z}_j \mathbf{z}_j^T \right)^{-1}.$$

That would provide no additional insights, however, to what follows, so from now on we will limit our attention to **g** with θ equal 0 or 1.

Whether θ is 0 or 1, t_{GREG} is a nearly unbiased estimator for T under *quasi-probability theory* which treats response as a second phase of Poisson (i.e., independent) random sampling so long as each unit's probability of response has the form:

$$\rho_k = 1 / \left(1 + \boldsymbol{\gamma}^T \mathbf{z}_k\right), \tag{2}$$

and **g** is a consistent estimator for $\boldsymbol{\gamma}$. Moreover, $t_{\text{GREG}} = \sum_R w_k y_k = \sum_R d_k p_k^{-1} y_k$, where $p_k = 1/(1 + \mathbf{g}^T \mathbf{z}_k)$, is a nearly unbiased estimator for T.

4 A Useful Form of Nonlinear Calibration Weighting

The problem with the response model in Eq. (2) is that ρ_k and its estimate p_k can be negative. A useful nonlinear form of calibration weighting that avoids this possibility finds a **g** (through repeated linearizations, i.e., Newton's method) such that

$$\sum_{k \in R} w_k \mathbf{z}_k = \sum_{k \in R} d_k \alpha \left(\mathbf{g}^T \mathbf{z}_k\right) \mathbf{z}_k = (1 - \theta) \sum_{k \in U} \mathbf{z}_k + \theta \sum_{k \in S} d_k \mathbf{z}_k, \tag{3}$$

where

$$\alpha \left(\mathbf{g}^T \mathbf{z}_k\right) = \frac{\ell + \exp\left(\mathbf{g}^T \mathbf{z}_k\right)}{1 + \exp\left(\mathbf{g}^T \mathbf{z}_k\right)/u}; \quad 0 \leq \ell < u \leq \infty, \tag{4}$$

and θ (again) is 0 or 1.

The *weight adjustment* $\alpha(\mathbf{g}^T \mathbf{z}_k)$ has a lower bound $\ell \geq 0$ and an upper bound $u > \ell$, which can be infinite. These *bounding* parameters force $1/u \leq p_k \leq 1/\ell$.

Equation (4) is a generalization of both raking, where $\ell = 0, u = \infty$, and the implicit estimation of a logistic-regression response model, $\rho_k = 1/[1 + \exp(\boldsymbol{\gamma}^T \mathbf{x}_k)]$, where $\ell = 1, u = \infty$, and **g** is a consistent estimator for $\boldsymbol{\gamma}$.

When $\ell < 1$, Eq. (4) is the generalized-raking adjustment introduced by Deville and Särndal (1992), which potentially bounds the range of the $\alpha(\mathbf{g}^T \mathbf{z}_k)$. This bounding happens when ℓ is positive or u is less than infinity. For nonresponse adjustment, however, it is more reasonable to assume $\ell \geq 1$. Note that when calibration weighting is used to adjust for frame coverage errors resulting from duplication, a subject we will not discuss further here, it is reasonable to allow ℓ to be less than 1.

When a component of $\mathbf{z}_k$ is a constant, Kott and Liao (2012) showed that Eq. (4) is a special case of Folsom and Singh's (2000) general exponential (weight-adjustment) model

$$\alpha_k \left(\mathbf{g}^T \mathbf{z}_k\right) = \frac{\ell_k \left(u_k - c_k\right) + u_k \left(c_k - \ell_k\right) \exp\left(A_k \mathbf{g}^T \mathbf{z}_k\right)}{\left(u_k - c_k\right) + \left(c_k - \ell_k\right) \exp\left(A_k \mathbf{g}^T \mathbf{z}_k\right)},$$

where $A = (u_k - \ell_k)/[(u_k - c_k)(c_k - \ell_k)]$, but with a common $\mathbf{g}$ chosen to satisfy the calibration equation in (3). Although the general exponential model allows $\alpha_k(\mathbf{g}^T \mathbf{z}_k)$ to be k-specific, when adjusting for nonresponse (or coverage), it is sensible to select a single value for the c_k and a very limited number of ℓ_k and u_k values.

5 Model Variables and Calibration Variables

Deville (2000) pointed out that calibration weighting can be used when nonrespondents were not missing at random. An example is the response model

$$\rho_k = \left[\alpha \left(\boldsymbol{\gamma}^T \mathbf{x}_k\right)\right]^{-1} = \frac{1 + \exp\left(\boldsymbol{\gamma}^T \mathbf{x}_k\right)/u}{\ell + \exp\left(\boldsymbol{\gamma}^T \mathbf{x}_k\right)},$$

where some components of the *model* vector $\mathbf{x}_k$ are known only for respondents (such as y_k itself) but the $\mathbf{z}$-vector in the slightly revised calibration equation

$$\sum_R d_k \alpha \left(\mathbf{g}^T \mathbf{x}_k\right) \mathbf{z}_k = (1 - \theta) \sum_U \mathbf{z}_k + \theta \sum_S d_k \mathbf{z}_k \tag{5}$$

contains values known for respondents and nonrespondents.

In Deville's formulation the model and calibration vectors, $\mathbf{x}_k$ and $\mathbf{z}_k$, have the same number of components. Chang and Kott (2008) relaxed this requirement, but only by redefining what a calibration estimator is. We leave that generalization for a later section.

Consider for a moment linear calibration to the population with $\alpha(\mathbf{g}^T \mathbf{x}_k) = 1 + \mathbf{g}^T \mathbf{x}_k$ and $\mathbf{x}_k \neq \mathbf{z}_k$. A modification of Eq. (1) still describes a calibration estimator for T, but S is replaced by R, while $\mathbf{b}_S$ is replaced by $\mathbf{b}_R = (\sum_R d_j \mathbf{x}_j \mathbf{z}_j^T)^{-1} \sum_R d_k \mathbf{x}_k y_k$. The latter looks very much like an instrumental-variable regression coefficient in econometrics. The role of the components of $\mathbf{x}_k$ as explanatory variables in the response model is, however, roughly opposite their traditional role of instrumental variables in econometrics, where under a prediction (outcome) model for y_k, $E(y_k \mid \mathbf{z}_k, \mathbf{x}_k) = \mathbf{z}_k^T \boldsymbol{\beta}$. The $\mathbf{x}_k$ have no direct role in this prediction model. Their use is purely in estimation—like the $\mathbf{z}_k$ in the response model.

Consequently, a better name for a response-model variable that is not a calibration variable is a "model-only variable." In the same spirit, a calibration variable that is not a response model variable could be called a "shadow variable." It is a shadow variable that functions like a traditional instrumental variable (in fact, Kim and Shao 2013, p. 126, use the term "nonresponse instrument" to describe such a variable). On the one hand, a good shadow variable will be correlated with at least one model-only variable. On the other, given the full set of response-model variables, a shadow variable's own value should be unrelated to the probability of unit response.

6 Two Models for Nonresponse Adjustment

Let I_k be a sample indicator (1 if $k \in S$, 0 otherwise), R_k be a response indicator (1 if k responds, 0 otherwise), w_k be a calibration weight, and p_k be a possibly bad implicit guess at $\rho_k = E(R_k)$ used in the weight $w_k = R_k(d_k/p_k)$. Note that $w_k = 0$ when $R_k = 0$.

The bias in a calibration estimator satisfying Eq. (5) with $\theta = 1$ can be expressed as

$$E\left(\sum_{k \in R} w_k y_k - T\right) = E\left(\sum_{k \in S} w_k R_k y_k - \sum_{k \in S} d_k y_k\right)$$

$$= -E\left(\sum_{k \in S} \frac{d_k}{p_k} y_k [p_k - R_k]\right) \tag{6}$$

$$= -E\left(\sum_{k \in S} \frac{d_k}{p_k} [y_k - \mathbf{z}_k^T \mathbf{b}][p_k - R_k]\right)$$

for *any* fixed vector $\mathbf{b}$.

When calibrating with $\theta = 0$, Eq. (6) becomes only a near equality, but these three conclusions remain:

1. The calibration weighted estimator is nearly unbiased under quasi-probability-sampling theory assuming the Poisson response model used to produce the calibration weights holds:

$$\rho_k = E(R_k) = p\left(\mathbf{\gamma}^T \mathbf{x}_k \middle| \{y_m, \mathbf{x}_m, \mathbf{z}_m, I_m; m \in U\}\right) = p\left(\mathbf{\gamma}^T \mathbf{x}_k\right). \tag{7}$$

2. The calibration weighted estimator is nearly unbiased under a combination of probability sampling and the following *prediction model*:

$$y_k = \mathbf{z}_k^T \mathbf{\beta} + \varepsilon_k, \tag{8}$$

where

$$E\left(\varepsilon_k \middle| \{\mathbf{z}_m, \mathbf{x}_m, I_m, R_m; m \in U\}\right) = 0,$$

$$E\left(\varepsilon_k^2 \middle| \{\mathbf{z}_m, \mathbf{x}_m, I_m, R_m; m \in U\}\right) = \sigma_k^2 < \infty.$$

No particular response model is assumed to hold, but p_k, which need not be a consistent estimator for ρ_k, remains a function of $\mathbf{z}_k$ and $\mathbf{x}_k$.
3. The calibration weighted estimator can be nearly unbiased [the right-hand side of Eq. (6) nearly zero] when *neither* of the models in the above two conclusions holds.

The property that the near unbiasedness of the calibration estimator obtains if *either* an assumed response or an assumed prediction model holds has been called "double protection again nonresponse bias" by Kim and Park (2006) and "double robustness" in the biostatistics literature (Bang and Robins 2005).

A simulation result in D'Arrigo and Skinner (2010) appears to provide an example that those model assumptions are sufficient but not necessary. In that simulation, total employment was estimated after nonresponse had been simulated with a raking model ($\rho_k = \exp(\boldsymbol{\gamma}^T \mathbf{z}_k)$). The response model was clearly not $\rho_k = 1/[1 + \exp(\boldsymbol{\gamma}^T \mathbf{z}_k)]$ nor was employment a linear function of the components of $\mathbf{z}_k$. Nevertheless, using linear calibration did not appear to result in a systematic bias.

Observe that the roles of the $\mathbf{x}_k$ and $\mathbf{z}_k$ vectors are reversed in the response and prediction models in Eqs. (7) and (8) respectively. This has a consequence. For a response model, there is no reason a y_k cannot be a component of $\mathbf{x}_k$. The same cannot be said about the prediction model at this point since it makes little sense to model $E(y_k | \mathbf{z}_k, y_k)$ as a function of $\mathbf{z}_k$. In a later section, we will expand the meaning of a prediction model.

7   The Response Model

In this section, we assume only a Poisson response model with a known form $\rho_k = p(\boldsymbol{\gamma}^T \mathbf{x}_k)$, where $p(\phi)$ is monotonic and twice differentiable. Assuming a Poisson response model means that if population element k is sampled, its probability of response (self-selection) has the form specified and is independent of whether other elements respond when sampled.

One can estimate the unknown parameter $\boldsymbol{\gamma}$ implicitly by finding the $\mathbf{g}$ that satisfies the calibration equation in (5) for $\theta = 0$ or 1. Under mild conditions on the sample design, population, and $p(\phi)$, as the original sample size n (and thus the respondent sample size) grows large, $\mathbf{g} - \boldsymbol{\gamma} = \mathbf{O}_P(1/\sqrt{n})$, and:

$$\alpha\left(\mathbf{g}^T \mathbf{x}_k\right) - \alpha\left(\boldsymbol{\gamma}^T \mathbf{x}_k\right) = \alpha'\left(\mathbf{g}^T \mathbf{x}_k\right)(\mathbf{g} - \boldsymbol{\gamma})^T \mathbf{x}_k + O_P\left(1/n\right), \qquad (9)$$

where $\alpha(\phi) = 1/p(\phi)$, and $\alpha'(\phi)$ is its derivative.

Let us define the *quasi-probability estimated regression coefficient* as

$$\mathbf{b}_R = \left(\sum_{k \in S} d_k R_k \alpha' \left(\mathbf{g}^T \mathbf{x}_k \right) \mathbf{x}_k \mathbf{z}_k{}^T \right)^{-1} \sum_{k \in S} d_k R_k \alpha' \left(\mathbf{g}^T \mathbf{x}_k \right) \mathbf{x}_k y_k$$

$$= \left(\mathbf{A}^T \right)^{-1} \sum_{k \in S} d_k R_k \alpha' \left(\mathbf{g}^T \mathbf{x}_k \right) \mathbf{x}_k y_k / N,$$

where $\mathbf{A}^T = \sum_S d_k R_k \alpha'(\mathbf{g}^T \mathbf{x}_k) \mathbf{x}_k \mathbf{z}_k{}^T / N$ is assumed to be invertible. Moreover, let the *quasi-probability residual* be $e_k = y_k - \mathbf{z}_k{}^T \mathbf{b}_R$, so that $\sum_S d_k R_k \alpha'(\mathbf{g}^T \mathbf{x}_k) \mathbf{x}_k$ $e_k = \sum_R d_k \alpha'(\mathbf{g}^T \mathbf{x}_k) \mathbf{x}_k e_k = 0$ by definition. It is for this property that $\alpha'(\mathbf{g}^T \mathbf{x}_k)$ has been injected into the quasi-probability estimated regression coefficient.

The error in a calibrated estimator for T is

$$t_{\text{CAL}} - T = \sum_{k \in S} w_k y_k - \sum_{k \in U} y_k$$

$$= \left(\sum_{k \in S} w_k \mathbf{z}_k{}^T \mathbf{b}_R + \sum_{k \in S} w_k e_k \right) - \left(\sum_{k \in U} \mathbf{z}_k{}^T \mathbf{b}_R + \sum_{k \in U} e_k \right)$$

$$= \left(\sum_{k \in S} w_k e_k - \sum_{k \in U} e_k \right) + \theta \left(\sum_{k \in S} d_k \mathbf{z}_k{}^T \mathbf{b}_R - \sum_{k \in U} \mathbf{z}_k{}^T \mathbf{b}_R \right)$$

$$= \sum_{k \in S} d_k R_k \alpha \left(\mathbf{\gamma}^T \mathbf{x}_k \right) e_k - \sum_{k \in S} d_k R_k \alpha \left(\mathbf{\gamma}^T \mathbf{x}_k \right) e_k$$

$$+ \left(\sum_{k \in S} d_k R_k \alpha \left(\mathbf{g}^T \mathbf{x}_k \right) e_k - \sum_{k \in U} e_k \right) + \theta \left(\sum_{k \in S} d_k \mathbf{z}_k{}^T \mathbf{b}_R - \sum_{k \in U} \mathbf{z}_k{}^T \mathbf{b}_R \right).$$

Dividing both sides by N, so that $O_P(1/n)$ terms can be dropped under mild conditions and using Eq. (9) produces:

$$(t_{\text{CAL}} - T) / N \approx \sum_{k \in S} d_k R_k \alpha' \left(\mathbf{g}^T \mathbf{x}_k \right) (\mathbf{g} - \mathbf{\gamma})^T \mathbf{x}_k e_k / N$$

$$+ \left(\sum_{k \in S} d_k R_k \alpha \left(\mathbf{\gamma}^T \mathbf{x}_k \right) e_k - \sum_{k \in U} e_k \right) / N$$

$$+ \theta \left(\sum_{k \in S} d_k \mathbf{z}_k{}^T \mathbf{b}_R - \sum_{k \in U} \mathbf{z}_k{}^T \mathbf{b}_R \right) / N$$

$$= (\mathbf{g} - \mathbf{\gamma})^T \sum_{k \in S} d_k R_k \alpha' \left(\mathbf{g}^T \mathbf{x}_k \right) \mathbf{x}_k e_k / N$$

$$+\left(\sum_{k\in S} d_k R_k \alpha \left(\boldsymbol{\gamma}^T \mathbf{x}_k\right) e_k - \sum_{k\in U} e_k\right)/N$$

$$+\theta\left(\sum_{k\in S} d_k \mathbf{z}_k{}^T \mathbf{b}_R - \sum_{k\in U} \mathbf{z}_k{}^T \mathbf{b}_R\right)/N$$

$$=\left(\sum_{k\in S} d_k R_k \alpha \left(\boldsymbol{\gamma}^T \mathbf{x}_k\right) e_k - \sum_{k\in U} e_k\right)/N$$

$$+\theta\left(\sum_{k\in S} d_k \mathbf{z}_k{}^T \mathbf{b}_R - \sum_{k\in U} \mathbf{z}_k{}^T \mathbf{b}_R\right)/N. \tag{10}$$

From Eq. (10), we can easily see that calibrating to the sample ($\theta = 1$) will usually produce a calibration estimator for T with more mean squared error than calibrating to the population ($\theta = 0$) since the former has an added term in its error: $\sum_S d_k \mathbf{z}_k{}^T \mathbf{b}_R - \sum_U \mathbf{z}_k{}^T \mathbf{b}_R$. This will certainly be true when one assumes the prediction model in Eq. (8) since $e_k \approx \varepsilon_k$ under mild conditions, and ε_k is uncorrelated with $\mathbf{z}_k$. Otherwise, it is only very likely to be true.

Suppose we have a stratified multistage sample design with H strata, n_h primary sampling units (PSUs) in stratum h, and ignorably small first-stage sampling fractions. Assuming the mild conditions under which the small bias of t_{CAL} is an asymptotically ignorable component of its mean squared error, Eq. (10) suggests that a good variance/mean squared error estimator for t_{CAL} is

$$v_{WR}\left(t_{CAL}\right) = \sum_{h=1}^{H} \frac{n_h}{n_h - 1} \sum_{j=1}^{n_h} \left\{\left(\sum_{k\in S_{hj}} \psi_k\right)^2 - \left(\frac{1}{n_h}\sum_{S_{hj}}\sum \psi_k\right)^2\right\}, \tag{11}$$

where $\psi_k = \theta d_k \mathbf{z}_k{}^T \mathbf{b}_R + w_k e_k$, and S_{hj} is the set of sampled elements in PSU j of stratum h.

A formal proof of the effectiveness of the linearization-based variance estimator in Eq. (11) (see, for example, Kott and Liao 2014) would need to recognize that the variance of t_{CAL} is asymptotically equivalent to the variance of

$$\left(\sum_{k\in S} d_k R_k \alpha \left(\boldsymbol{\gamma}^T \mathbf{x}_k\right) e_k* - \sum_{k\in U} e_k*\right) + \theta\left(\sum_{k\in S} d_k \mathbf{z}_k{}^T \mathbf{b}* - \sum_{k\in U} \mathbf{z}_k{}^T \mathbf{b}*\right),$$

where $\mathbf{b}_R*$, the probability limit of $\mathbf{b}_R$, is assumed to exist. Neither $\mathbf{b}_R*$ nor $e_k* = y_k - \mathbf{z}_k{}^T \mathbf{b}_R*$ depend on the realized respondent sample. Although

$$(\mathbf{g} - \boldsymbol{\gamma})^T \sum_{k\in S} d_k R_k \alpha' \left(\mathbf{g}^T \mathbf{x}_k\right) \mathbf{x}_k e_k * /N$$

is not zero like $(\mathbf{g}-\boldsymbol{\gamma})^T \sum s d_k R_k \alpha'(\mathbf{g}^T \mathbf{x}_k) \mathbf{x}_k e_k/N$, it is $O_P(1/n)$ under mild conditions and thus ignorably small. Finally, the $\psi_k = \theta d_k \mathbf{z}_k^T \mathbf{b}_R + w_k e_k = d_k$ $(\mathbf{z}_k^T \mathbf{b}_R + [R_k/p_k]e_k)$ terms in Eq. (11) capture nearly all of the possible correlation between $\sum s \theta d_k \mathbf{z}_k^T \mathbf{b}_R *$ and $\sum s w_k e_k *$.

8 Adjusting for Nonresponse by Fitting a Logistic Response Function Directly

A popular alternative to calibrating to the sample is to fit a logistic response model using maximum-likelihood (ML) or weighted maximum-likelihood (WML). In other words, first find a $\mathbf{g}$ such that $\sum s \lambda_k [R_k - p(\mathbf{g}^T \mathbf{x}_k)]\mathbf{x}_k = \mathbf{0}$, where $\lambda_k = 1$ when using ML or d_k when using WML, and then compute the weights: $w_k = R_k(d_k/p_k)$ for use in either $t_{\text{ML}} = \sum s w_k y_k$ or $t_{\text{WML}} = \sum s w_k y_k$.

This is the same as solving for

$$\sum_{k \in S} \frac{d_k \left[R_k - p\left(\mathbf{g}^T \mathbf{x}_k\right)\right] \lambda_k}{p\left(\mathbf{g}^T \mathbf{x}_k\right)} \frac{\lambda_k}{d_k} p\left(\mathbf{g}^T \mathbf{x}_k\right) \mathbf{x}_k = \mathbf{0}$$

or

$$\sum_{k \in S} d_k \left[\frac{R_k}{p\left(\mathbf{g}^T \mathbf{x}_k\right)} - 1 \right] \left(\frac{\lambda_k}{d_k} p_k \mathbf{x}_k \right) = \mathbf{0},$$

which in turn is equivalent to calibrating to the sample with $\alpha(\mathbf{g}^T \mathbf{x}_k) = 1 + \exp(\mathbf{g}^T \mathbf{x}_k)$, and $\mathbf{z}_k = (\lambda_k/d_k)p_k\mathbf{x}_k$.

This realization is not directly useful since the components of $\mathbf{z}_k$ are unknown ex ante because p_k is unknown. It does, however, suggest that the variance estimator in Eq. (11) can be employed when nonresponse is adjusted using a fitted logistic response model as described above. One needs, of course, assume that the response is indeed logistic in $\mathbf{x}_k$ and independent across elements.

There is no combined probability-sampling/prediction-model justification for t_{ML} or t_{WML} as there is for t_{CAL} with $\alpha(\mathbf{g}^T \mathbf{z}_k) = 1 + \exp(\mathbf{g}^T \mathbf{z}_k)$. This was noted by Kim and Riddles (2012), who argued that calibration weighting is often more efficient than ML and WML when the goal is estimating a population total (or mean) not the nonresponse parameter $\boldsymbol{\gamma}$ per se.

9 The Variance of g

Assuming an arbitrary Poisson response model holds, we can plug equation (9) into calibration equation (5) and rearrange terms to get an expression for the error in $\mathbf{g}$ as an estimator for $\boldsymbol{\gamma}$:

$$\mathbf{g} - \boldsymbol{\gamma} \approx -\mathbf{A}^{-1}\left[\sum_{k\in S} d_k R_k \alpha\left(\boldsymbol{\gamma}^T \mathbf{x}_k\right)\mathbf{z}_k - \left((1-\theta)\sum_{k\in U}\mathbf{z}_k + \theta\sum_{k\in S} d_k\mathbf{z}_k\right)\right],$$

where $\mathbf{A} = N^{-1}\sum_S d_k R_k \alpha'(\mathbf{g}^T\mathbf{x}_k)\mathbf{z}_k\mathbf{x}_k^T$.

Letting $\mathbf{A}^*$ be the probability limit of $\mathbf{A}$, the variance of $\mathbf{g}$ is then

$$\mathrm{Var}(\mathbf{g}) \approx \mathbf{A}^{*-1}\mathrm{Var}_{I,R}\left[\frac{1}{N}\sum_S d_k\left(R_k/\rho_k\right)\mathbf{z}_k\right]\left(\mathbf{A}^{*-1}\right)^T,$$

where $\rho_k = 1/\alpha(\boldsymbol{\gamma}^T\mathbf{x}_k)$, and $E(R_k) = \rho_k$. When calibrating to the sample,

$$\mathrm{Var}_{I,R}\left[\frac{1}{N}\sum_{k\in S} d_k\left(R_k/\rho_k\right)\mathbf{z}_k\right] = \mathrm{Var}_R\left[\frac{1}{N}\sum_{k\in S} d_k\left(R_k/\rho_k\right)\mathbf{z}_k\Big| S\right]$$

$$= \frac{1}{N^2}\sum_{k\in S} d_k{}^2\left(1/\rho_k\right)\left(1-\rho_k\right)\mathbf{z}_k\mathbf{z}_k{}^T,$$

which can be estimated by $N^{-2}\sum_S d_k{}^2 R_k(1/p_k{}^2)(1-p_k)\mathbf{z}_k\mathbf{z}_k{}^T$, while $\mathbf{A}^*$ is estimated by $\mathbf{A}$.

When calibrating to the population, there is an additional term, the variance of $\sum_S d_k\mathbf{z}_k$ under probability sampling theory. A more direct estimator for the variance of $\mathbf{g}$ under the stratified multistage design described in Sect. 7 is

$$\mathbf{v}_{\mathrm{WR}}(\mathbf{g}) = \mathbf{A}\sum_{h=1}^{H}\frac{n_h}{n_h-1}\sum_{j=1}^{n_h}\left\{\left(\sum_{k\in S_{hj}} w_k\mathbf{z}_k\right)^2 - \frac{1}{n_h}\left(\sum_{S_{hj}}\sum w_k\mathbf{z}_k\right)^2\right\}\mathbf{A}^{-1},$$

where $\mathbf{q}^2 = \mathbf{q}\mathbf{q}^T$.

10 The Prediction Model

Let us turn to the prediction model:

$$y_k = \mathbf{z}_k{}^T\boldsymbol{\beta} + \varepsilon_k, \tag{8}$$

where

$$E\left(\varepsilon_k\Big| \{\mathbf{z}_m, \mathbf{x}_m, I_m, R_m; m\in U\}\right) = 0,$$

$$E\left(\varepsilon_k{}^2\Big| \{\mathbf{z}_m, \mathbf{x}_m, I_m, R_m; m\in U\}\right) = \sigma_k{}^2 < \infty,$$

and the ε_k are uncorrelated across PSUs. It is not hard to see that under the sampling design described in Sect. 7, Eq. (11) provides a nearly unbiased estimator for the prediction-model variance of t_{CAL} as an estimator for T when calibrating to the population, $\sum^{H}\sum^{n_h}\mathrm{Var}_\varepsilon\left(\sum_{S_{hj}} w_k\varepsilon_k\right)$. When calibrating to the sample, Eq. (11) provides a nearly unbiased estimator for the combination of the prediction-model variance and the variance under probability sampling theory of $\Sigma_S d_k z_k{}^T \boldsymbol{\beta}$ whatever the response model.

When $E(\varepsilon_k|\{\mathbf{z}_m, \mathbf{x}_m, I_m, R_m; m \in U\})$ does not equal 0 because, for example, the variable of interest is a component of $\mathbf{x}_m$, then Kott and Chang (2010) replaced the prediction model in (8) with a two-equation model:

$$y_k = \mathbf{x}_k{}^T \boldsymbol{\beta}_\mathbf{x} + \varepsilon_k, \quad \text{and} \quad \mathbf{z}_k{}^T = \mathbf{x}_k{}^T \boldsymbol{\Gamma} + \boldsymbol{\eta}_k{}^T, \tag{12}$$

where

$$E\left(\varepsilon_k \,\middle|\, \{\mathbf{x}_m, I_m, R_m; m \in U\}\right) = 0,$$

$$E\left(\varepsilon_k{}^2 \,\middle|\, \{\mathbf{x}_m, I_m, R_m; m \in U\}\right) = \sigma_k{}^2 < \infty,$$

$$E\left(\boldsymbol{\eta}_k \,\middle|\, \{\mathbf{x}_m, I_m, R_m; m \in U\}\right) = \mathbf{0},$$

$E(\boldsymbol{\eta}_k\boldsymbol{\eta}_k{}^T|\{\mathbf{x}_m, I_m, R_m; m \in U\}) = \sum_{\eta k}$ is positive definite, and $\boldsymbol{\eta}_k{}^+ = (\varepsilon_k, \boldsymbol{\eta}_k{}^T)^T$ is independent across PSUs.

Observe that if $\boldsymbol{\Gamma}$ is invertible, and $\mathbf{g}$ has a limit as the sample size grows arbitrarily large (recall we are not assuming a response model), then

$$\mathbf{b}_R{}^* = \left[\plim_{n\to\infty}\left(\frac{1}{N}\sum_S d_k R_k \alpha'\left(\mathbf{g}^T\mathbf{x}_k\right)\mathbf{x}_k\mathbf{z}_k{}^T\right)\right]^{-1}$$

$$\plim_{n\to\infty}\left(\frac{1}{N}\sum_S d_k R_k \alpha'\left(\mathbf{g}^T\mathbf{x}_k\right)\mathbf{x}_k y_k\right)$$

$$= \left[\plim_{n\to\infty}\left(\frac{1}{N}\sum_S d_k R_k \alpha'\left(\mathbf{g}^T\mathbf{x}_k\right)\mathbf{x}_k\left(\mathbf{x}_k{}^T\boldsymbol{\Gamma} + \boldsymbol{\eta}_k{}^T\right)\right)\right]^{-1}$$

$$\plim_{n\to\infty}\left(\frac{1}{N}\sum_S d_k R_k \alpha'\left(\mathbf{g}^T\mathbf{x}_k\right)\mathbf{x}_k\left(\mathbf{x}_k{}^T\boldsymbol{\beta}_\mathbf{x} + \varepsilon_k\right)\right)$$

$$= \boldsymbol{\Gamma}^{-1}\boldsymbol{\beta}_\mathbf{x},$$

and $e_k{}^* \big| \{\mathbf{x}_m\} = \left(y_k - \mathbf{z}_k{}^T \mathbf{b}^*\right) \big| \{\mathbf{x}_m\}$

$$= \left[\left(\mathbf{x}_k{}^T \boldsymbol{\beta}_\mathbf{x} + \varepsilon_k\right) - \left(\mathbf{x}_k{}^T \boldsymbol{\Gamma} + \boldsymbol{\eta}_k{}^T\right) \boldsymbol{\Gamma}^{-1} \boldsymbol{\beta}_\mathbf{x}\right] \Big| \{\mathbf{x}_m\}$$

$$= \left(\varepsilon_k - \boldsymbol{\eta}_k{}^T \boldsymbol{\Gamma}^{-1} \boldsymbol{\beta}_\mathbf{x}\right) \big| \{\mathbf{x}_m\}.$$

Let us call this last expression, ξ_k. Observe that when y_k is a component of $\mathbf{x}_k$, $\varepsilon_k = 0$, and all the error in t_{CAL} comes from calibrating on the $\mathbf{z}_k$ rather than the unknown $\mathbf{x}_k$.

Even when calibrating to the population, t_{CAL} is not a strictly prediction-model unbiased estimator for T given the $\mathbf{x}_k$ because the $(\mathbf{g}^T \mathbf{x}_k)$ term in $t_{\text{CAL}} = \Sigma_S d_k R_k \alpha (\mathbf{g}^T \mathbf{x}_k) y_k$ is not a constant since $\mathbf{g}$ depends on the $\mathbf{z}_k$ through the calibration equation. The estimator t_{CAL} is, however, *nearly* prediction-model unbiased under mild conditions since $\mathbf{g}$ does not depend on the $\mathbf{z}_k$ at its asymptotic limit. Moreover, the prediction model variance of t_{CAL} is nearly $\sum^H \sum^{n_h} \text{Var}\left(\sum_{S_{hj}} w_k \xi_k\right)$, and Eq. (11) estimates the calibration estimators prediction variance (when $\theta = 1$) or its combined prediction and probability-sampling variance under mild conditions.

From a prediction-model viewpoint, there is no reason to use a nonlinear form of calibration weighting rather than a linear form. When $\mathbf{z}_k = (1 \; z_k)^T$ and $\mathbf{x}_k = (1 \; y_k)^T$ are under linear calibration, we have the reverse regression proposed by Andridge and Little (2011) when nonresponse is exclusively a function of the survey variable.

11   When $\mathbf{z}_k$ Has More Components Than $\mathbf{x}_k$

Chang and Kott (2008) allowed the number of components in $\mathbf{z}_k$ to exceed the number of components in $\mathbf{x}_k$ even though this would make it nearly impossible to satisfy the calibration equation. Instead of finding a $\mathbf{g}$ such that

$$\mathbf{s} = \sum_{k \in S} d_k R_k \alpha \left(\mathbf{g}^T \mathbf{x}_k\right) \mathbf{z}_k - \left[(1 - \theta) \sum_{k \in U} \mathbf{z}_k + \theta \sum_{k \in S} d_k \mathbf{z}_k\right] = \mathbf{0}, \qquad (13)$$

they suggested minimizing $\mathbf{s}^T \mathbf{Q} \mathbf{s}$ for some invertible $\mathbf{Q}$ by choosing the $\mathbf{g}$ that solves $(\mathbf{s}')^T \mathbf{Q} \mathbf{s} = \mathbf{0}$, where $(\mathbf{s}')^T = \Sigma_S d_j R_j \alpha'(\mathbf{g}^T \mathbf{x}_j) \mathbf{x}_j \mathbf{z}_j{}^T$, that is, the transpose of the derivate of $\mathbf{s}$ with respect to $\mathbf{g}$.

This strategy is the same as solving Eq. (13) after replacing $\mathbf{z}_k$ with

$$\tilde{\mathbf{z}}_k = N^{-1} (\mathbf{s}')^T \mathbf{Q} \mathbf{z}_k. \qquad (14)$$

Most of our previous results repeat with this substitution, because $\tilde{\mathbf{z}}_k$ has the same number of components as $\mathbf{x}_k$. For many of the proofs $\tilde{\mathbf{z}}_k$ is replaced by its limit, which is assumed to exist [and which is why N^{-1} was inserting into the right-side side of Eq. (14)].

Chang and Kott showed that in the "degenerate" case where $\mathbf{z}_k$ has the same number of components as $\mathbf{x}_k$, the choice of $\mathbf{Q}$ didn't affect the resulting t_{CAL} (assuming $N^{-1}(\mathbf{s}')^T\mathbf{Q} = \mathbf{A}^T\mathbf{Q}$ is invertible). Otherwise, they argued that the best choice for $\mathbf{Q}$ when calibrating to the population would be

$$\mathbf{Q}^* = \left[\text{Var}_{I,R} \left(N^{-1} \sum_{k \in S} d_k \frac{R_k}{p\left(\boldsymbol{\gamma}^T \mathbf{x}_k\right)} \mathbf{z}_k \right) \right]^{-1}.$$

In order to determine $\mathbf{g}$ and $\mathbf{Q}^*$ simultaneously, they suggested an iterative routine starting with $\mathbf{g}^{(0)} = \mathbf{0}$. At iteration $r+1$, $\mathbf{Q}^{(r+1)}$ would be set equal to the estimated inverse of

$$\text{Var}_{I,R} \left(N^{-1} \sum_{k \in S} d_k \frac{R_k}{p\left(\mathbf{g}^{(r)T} \mathbf{x}_k\right)} \mathbf{z}_k \middle| \boldsymbol{\gamma} = \mathbf{g}^{(r)} \right),$$

computed as if $\mathbf{g}^{(r)}$ were the true value of $\boldsymbol{\gamma}$. After which, solving $(\mathbf{s}'^{(r+1)})^T \mathbf{Q}^{(r+1)}\mathbf{s}^{(r+1)} = 0$ would produce $\mathbf{g}^{(r+1)}$. They laid out the sufficient conditions for the convergence of this process.

Using their reasoning, when calibrating to the sample, $\mathbf{g}$ and $\mathbf{Q}$ would be iteratively updated using the inverse of

$$\text{Var}_R \left(N^{-1} \sum_{k \in S} d_k \frac{R_k}{p\left(\mathbf{g}^{(r)}\mathbf{x}_k\right)} \mathbf{z}_k \middle| \boldsymbol{\gamma} = \mathbf{g}^{(r)}; S \right) = \sum_{k \in S} d_k{}^2 \frac{1}{p_k} \left(1 - p_k\right) \mathbf{z}_k \mathbf{z}_k{}^T$$

or its estimate:

$$\mathbf{V}_{\mathbf{z}|\boldsymbol{\gamma}=\mathbf{g}^{(r)};S} = N^{-1} \sum_{k \in S} d_k{}^2 \frac{R_k}{p_k{}^2} \left(1 - p_k\right) \mathbf{z}_k \mathbf{z}_k{}^T$$

at step $r+1$.

An intriguing alternative approach to setting $\mathbf{Q}$ again uses an iterative process starting with $\mathbf{g}^{(0)} = \mathbf{0}$, but now

$$\mathbf{Q}^{(r+1)} = \left(N^{-1} \sum_{j \in S} d_j R_j \alpha' \left(\mathbf{g}^{(r)T} \mathbf{x}_j\right) \mathbf{z}_j \mathbf{z}_j{}^T \right)^{-1}.$$

Observe that if there were a matrix $\mathbf{D}$ such that $\mathbf{x}_k = \mathbf{D}\mathbf{z}_k$, then $\tilde{\mathbf{z}}_k$ in Eq. (14) would collapse into $\mathbf{x}_k$ since at every iteration (dropping the step-number superscripts for convenience):

$$\tilde{\mathbf{z}}_k = N^{-1}(\mathbf{s}')^T \mathbf{Q} \mathbf{z}_k$$

$$= \sum_{j \in S} d_j R_j \alpha' \left(\mathbf{g}^T \mathbf{x}_j \right) \mathbf{x}_j \mathbf{z}_j{}^T \left(\sum_{j \in S} d_j R_j \alpha' \left(\mathbf{g}^T \mathbf{x}_j \right) \mathbf{z}_j \mathbf{z}_j{}^T \right)^{-1} \mathbf{z}_k \qquad (15)$$

$$= \mathbf{B}_{(\mathbf{g})}{}^T \mathbf{z}_k,$$

where $\mathbf{B}_{(\mathbf{g})}{}^T = \mathbf{D}$ when $\mathbf{x}_k = \mathbf{D}\mathbf{z}_k$. More generally $\mathbf{B}_{(\mathbf{g})}$ is the estimated regression coefficient of $\mathbf{x}_k$ on $\mathbf{z}_k$ computed in the sample using the weights $d_j R_j \alpha'(\mathbf{g}^T \mathbf{x}_j)$.

It is very unlikely that such a $\mathbf{D}$ exists in practice, but if one did, then this is what we would want to happen. Under the prediction model in Eq. (12), $\boldsymbol{\eta}_k$ would be $\mathbf{0}$ and thus its contribution to ξ_k would disappear. When no such $\mathbf{D}$ exists, the linear regression in Eq. (15) forces $\tilde{\mathbf{z}}_k$ to be as close as possible to $\mathbf{x}_k$ in some sense, which would tend to reduce the contribution of $\boldsymbol{\eta}_k$ to ξ_k.

In their strictly prediction-model-based framework, Andridge and Little (2011) suggested something very close to Eq. (15) when $\mathbf{x}_k = (1 \ y_k)^T$ and $\mathbf{z}_k$ had more than two components. In their formulation, $d_j R_j \alpha'(\mathbf{g}^T \mathbf{x}_j)$ was replaced by R_j, and there was no need for iteration. Since their framework assumes the prediction model is correct, there would be no gain from finding the $\mathbf{g}$ that minimizes $\mathbf{s}^T \mathbf{Q} \mathbf{s}$ for some $\mathbf{Q}$.

12   Conclusions

Andridge and Little (2011) assumed y_k and $\mathbf{z}_k$ were multivariate normal. This explains how $\mathbf{z}_k$ could be a model of y_k as described in the second part of Eq. (12) when, say, y_k refers to a current-year variable while the components of $\mathbf{z}_k$ are from a previous year. At the very least, we need an appreciation that the prediction model is not a causal model. In fact, it is little more than a useful approximation of reality. As such, the fewer unnecessary assumptions, the better.

Under certain assumptions, we have seen that when there is unit nonresponse, calibrating to the population tends to be more statistically efficient than calibrating to the sample when estimating a population total, but the reverse is true when estimating the parameter of the assumed response model. Often in practice, one can do both: calibrate the first to the sample to adjust for nonresponse and then calibrate again to the population to increase statistical efficiency. See, for example, Kott and Liao (2014). This double adjustment makes variance estimation a bit more challenging unless replication is used.

A linearization-based variance estimator discussed in Kott and Liao (2014) can be used even when (first-stage) sampling fractions are not ignorably small, making replication ineffective. It assumes all the joint selection probabilities are bounded from below by a positive number, which rules out many systematic-sampling designs. Kott and Liao's variance estimator applies when assuming the response model in Eq. (7), or, after making additional assumptions about the element-level variances, when assuming the prediction model in Eq. (8). Kott and Liao did not discuss situations where the prediction model in Eq. (8) needed to be replaced by the two-equation model in (12).

Under a stratified, multistage sample with ignorably small first-stage selection probabilities, sufficient conditions on the sampling design for many of the results given here are that the selection probabilities be bounded from below by a positive number and the number of elements in a PSU be bounded as the sample size—and thus the number of PSUs—grows arbitrarily large. Also needed are restrictions on $1/p_k$, y_k, and the components of $\mathbf{z}_k$ and $\mathbf{x}_k$. Sufficient is that all be bounded from above by a finite number, while the population means of y_k, and the components of $\mathbf{z}_k$ and $\mathbf{x}_k$ be bounded from below by a positive number. That $1/p_k$, y_k, and the components of $\mathbf{z}_k$ and $\mathbf{x}_k$ be bounded from above seems reasonable given that we live in a finite world, but is actually quite strong when the sample size itself grows arbitrarily large. Alternative assumptions replace the finite upper bound on individual values with upper bounds on population moments.

One assumption that we did not label as mild is that $\mathbf{A} = N^{-1} \sum_s d_k R_k \alpha' (\mathbf{g}^T \mathbf{x}_k)$ $\mathbf{z}_k \mathbf{x}_k^T$ and its limit be invertible when $\mathbf{x}_k$ and $\mathbf{z}_k$ have the same number of components. This assumption can fail when there are model-only variables in $\mathbf{x}_k$ and shadow variables in $\mathbf{z}_k$. Without there being some linear correlation between these two sets, calibration weighting would not be a viable option. In practice, the calibration estimator t_{CAL} will not be very efficient when this correlation is weak. Moreover, assuming the response model is correctly specified, a weak correlation between shadow and model-only variables will result in an inefficient implicit estimate of $\boldsymbol{\gamma}$.

Worse, when response is partially a function of what are incorrectly assumed to be shadow variables, t_{CAL} can be badly biased. See Lesage and Haziza (2014). More work is needed on strategies for choosing model and calibration variables. For now, conducting sensitivity analyses under a variety of alternative models and calibration schemes is a prudent policy.

Finally, the WTADJX procedure in SUDAAN 11® (RTI 2012) allows a user to implement calibrating weighting given a Folsom–Singh weight adjustment function,

$$\alpha_k \left(\mathbf{g}^T \mathbf{x}_k\right) = \frac{\ell_k \left(u_k - c_k\right) + u_k \left(c_k - \ell_k\right) \exp\left(A_k \mathbf{g}^T \mathbf{x}_k\right)}{\left(u_k - c_k\right) + \left(c_k - \ell_k\right) \exp\left(A_k \mathbf{g}^T \mathbf{x}_k\right)},$$

and a calibration vector $\mathbf{z}_k$ that can have more components than $\mathbf{x}_k$ [with limited choices for $\mathbf{Q}$ in Eq. (14)]. The user supplies the bounding parameters in this adjustments function. When there is a single calibration-weighting step, the procedure can produce linearization-based standard error estimates under probability or quasi-probability sampling theory.

References

Andridge, R.R., Little, R.J.A.: Proxy pattern-mixture analysis for survey nonresponse. J. Off. Stat. **27**, 153–180 (2011)

Bang, H., Robins, J.M.: Doubly robust estimation in missing data and causal inference models. Biometrics **61**, 962–972 (2005)

Chang, T., Kott, P.S.: Using calibration weighting to adjust for nonresponse under a plausible model. Biometrika **95**, 557–571 (2008)

D'Arrigo, J., Skinner, C.J.: Linearization variance estimation for generalized raking estimators in the presence of nonresponse. Surv. Methodol. **36**, 181–192 (2010)

Deville, J.C., Särndal, C.E.: Calibration estimators in survey sampling. J. Am. Stat. Assoc. **87**, 376–382 (1992)

Deville, J.C.: Generalized calibration and application to weighting for non-response. In: COMP-STAT: Proceedings in Computational Statistics, 14th Symposium, Utrecht, Physica Verlag, Heidelberg (2000)

Folsom, R.E.: Exponential and logistic weight adjustments for sampling and nonresponse error reduction. In: Proceedings of the American Statistical Association, Social Statistics Section, pp. 197–202 (1991)

Folsom, R.E., Singh, A.C.: The generalized exponential model for sampling weight calibration for extreme values, nonresponse, and poststratification. In: Proceedings of the American Statistical Association, Survey Research Methods Section, pp. 598–603 (2000)

Fuller, W.A.: Sampling Statistics. Wiley, Hoboken (2009)

Fuller, W.A., Loughin, M.M., Baker, H.D.: Regression weighting for the 1987–88 National Food Consumption Survey. Surv. Methodol. **20**, 75–85 (1994)

Kim, J.K., Park, H.: Imputation using response probability. Can. J. Stat. **34**, 1–12 (2006)

Kim, J.K., Riddles, M.: Some theory for propensity scoring adjustment estimator. Surv. Methodol. **38**, 157–165 (2012)

Kim, J.K., Shao, J.: Statistical Methods for Handling Incomplete Data. Chapman and Hall/CRC, London (2013)

Kott, P.S.: Using calibration weighting to adjust for non-response and coverage errors. Surv. Methodol. **32**, 133–142 (2006)

Kott, P.S.: Calibration weighting: combining probability samples and linear prediction models. In: Pfeffermann, D., Rao, C.R. (eds.) Handbook of Statistics: Sample Surveys: Inference and Analysis, vol. 29B. Elsevier, New York (2009)

Kott, P.S., Chang, T.C.: Using calibration weighting to adjust for nonignorable unit nonresponse. J. Am. Stat. Assoc. **105**, 1265–1275 (2010)

Kott, P.S., Liao, D.: Providing double protection for unit nonresponse with a nonlinear calibration-weighting routine. Surv. Res. Methods **6**, 105–111 (2012)

Kott, P.S., Liao, D.: One step or two? Calibration weighting from a complete list frame with nonresponse. Surv. Methodol. (2014, forthcoming)

Lesage, E., Haziza, D.: On the problem of bias and variance amplification of the instrumental calibration estimator in the presence of unit nonresponse. J. Surv. Stat. Methodol. (2014) (forthcoming)

Lundström, S., Särndal, C.E.: Calibration as a standard method for treatment of nonresponse. J. Off. Stat. **15**, 305–327 (1999)

RTI: *SUDAAN Language Manual, Release 11.0.* RTI International, Research Triangle Park, NC (2012)

Rubin, D.B.: Inference and missing data (with discussion). Biometrika **63**, 581–592 (1976)

M-Quantile Small Area Models for Measuring Poverty at a Local Level

Monica Pratesi

Abstract M-quantile small area estimation (SAE) methods constitute a set of advanced statistical inference techniques that can be used for the measurement of poverty and living conditions by survey practitioners, researchers in private and public organizations, official statistical agencies, and local governmental agencies. In particular, the estimates produced using these SAE methods are well suited to mapping geographical variations in these conditions. In this paper, we summarize the ideas set out in some recent papers on M-quantile methods and their extensions and also comment on important issues that arise when SAE methods are used in poverty assessment in three Italian Regions.

Keywords M-quantile models • Poverty mapping • Small area estimation

1 Introduction

Over the last decade, there has been growing demand from both public and private sectors for measuring poverty and living conditions at disaggregated geographical levels, often referred to as small areas or small domains (Rao 2003). This increasing request of small area statistics is due, among other things, to their growing use in formulating policies and programs, in the allocation of government funds and in regional planning. Censuses provide "total" information, but only on a limited number of characteristics and once every 10 years. Statistical surveys produce high quantities of data and estimates, but cost constraints in the design of sample surveys lead to small sample sizes within small areas. As a result, direct estimation using only the survey data is inappropriate as it yields estimates with unacceptable levels

M. Pratesi (✉)
Department of Economics and Management, University of Pisa, via C. Ridolfi 10, 56124 Pisa, Italy
e-mail: m.pratesi@ec.unipi.it

F. Mecatti et al. (eds.), *Contributions to Sampling Statistics*, Contributions to Statistics, DOI 10.1007/978-3-319-05320-2__2,
© Springer International Publishing Switzerland 2014

of precision. In such cases small area estimation (SAE) can be performed via models that "borrow strength" by using all the available data and not only the area-specific data.

M-quantile SAE is about combining survey data that measures the target variables with auxiliary information about the population of interest. At its heart, the approach—proposed by Chambers and Tzavidis (2006)—models the M-quantiles of y given X rather than the expected value of y as would be the case with a standard regression model, and then characterizes the differences between the small areas using different M-quantile orders, and hence different fitted models.

M-quantile SAE models can be applied in many areas of statistical research: environmental statistics, economics, demography, epidemiology, and so on. Every study shows that the use of M-quantile modeling can produce reliable estimates, sometimes more accurate than that obtained by traditional methods. The first studies that connect poverty indicators and M-quantile methods are described in the studies by Tzavidis et al. (2008) and Pratesi et al. (2011). In the following years, many papers have been published showing how a tailored specification of the model and the use of auxiliary information improves the estimation of the small area parameters, both increasing efficiency and reducing bias (Chambers and Pratesi 2013).

In this paper, in Sect. 2 we review the M-quantile linear models that include area effects to account for between area variation beyond that explained by auxiliary variables and in Sect. 3 we focus on poverty mapping of the generalized measures of poverty introduced by Foster, Greer, and Thorbecke (the so-called FGT measures) in Italy. We refer, among others, to Giusti et al. (2011), Pratesi et al. (2009), and Giusti et al. (2009). Additional results obtained by these methods are available and downloadable from the web site of the SAMPLE project, funded by the European Commission under the 7FP (http://www.sample-project.eu). In Sect. 4 we extend these ideas to nonparametric M-quantile regression models (Pratesi et al. 2008a), and also to SAE techniques that use spatial information. Particularly, we illustrate an extension of M-quantile approach, where the nonstationarity in the data is captured via geographically weighted regression (GWR) (Salvati et al. 2013; Salvati et al. 2012). Our final remarks and the lessons learned from the applications of the models are described in the same section.

2   The M-Quantile Approach to Local Poverty Estimates

Among local poverty indicators the so-called Laeken indicators are used to target poverty and inequalities for comparisons between regions and countries. They are a core set of statistical indicators on poverty and social exclusion agreed by the European Council in December 2001, in the Brussels suburb of Laeken, Belgium. Here we focus on the estimation of the small area mean income and on the estimation of the incidence of poverty or *Head Count Ratio* (HCR) and the *Poverty*

Gap (PG), as denoted in the generalized measures of poverty introduced by Foster et al. (1984).

The population small area mean can be written as

$$m_d = N_d^{-1} \left(\sum_{j \in s_d} y_{jd} + \sum_{j \in r_d} y_{jd} \right), \tag{1}$$

where r_d denotes the $N_d - n_d$ non-sampled units while s_d the n_d sampled units in area d. Since the y values for the r_d non-sampled units are unknown, they need to be predicted.

Under the M-quantile setting, the small area mean estimator is obtained using the distribution function estimator due to Chambers and Dustan (1996) (CD hereafter). The MQ/CD estimator of the small area mean is

$$m_d^{\text{MQ/CD}} = N_d^{-1} \left\{ \sum_{j \in s_d} y_{jd} + \sum_{j \in r_d} \mathbf{x}_{jd}^{\text{T}} \hat{\boldsymbol{\beta}}_\psi (\hat{\theta}_d) \right.$$

$$\left. + \frac{N_d - n_d}{n_d} \sum_{j \in s_d} [y_{jd} - \mathbf{x}_{jd}^{\text{T}} \hat{\boldsymbol{\beta}}_\psi (\hat{\theta}_d)] \right\}, \tag{2}$$

which is based on the linear M-quantile model

$$y_{jd} = \mathbf{x}_{jd}^{\text{T}} \boldsymbol{\beta}_\psi (q_{jd}) + \varepsilon_{jd}. \tag{3}$$

Here ψ is the influence function associated with the qth M-quantile and the estimate $\hat{\boldsymbol{\beta}}_\psi (q_{jd})$ of $\boldsymbol{\beta}_\psi (q_{jd})$ is obtained, for specified q and continuous ψ, via an iterative weighted least squares algorithm. In estimator (2), $\hat{\theta}_d$ is an estimate of the average value of the M-quantile coefficients of the units in area d. Usually the influence function is the Huber proposal 2 and ε_{jd} has a nonspecified distribution.

M-quantile models automatically provide outlier robust inference. This inbuilt robustness resides in the expression of the residual adjustment in (2) but also in the no need to rely on the distributional assumptions that are inherent in mixed models, the models traditionally used for SAE (Rao 2003). This is a practical advantage when the outcome variable has a skewed distribution that can be affected by outliers, as it often happens in studies on income, poverty, and deprivation.

The flexibility and simplicity of M-quantile modeling resides also in the no need to specify a model for the random effects. In poverty mapping we start from the idea that the geographical distribution of the indicators is variable by areas and that the heterogeneity and variability cannot be completely captured by the covariates of the small area model. The area effect can be considered random in the sense that it is the area-specific residual reflecting remaining between area variability. In more details, traditional small area model-based estimators borrow strength from all the sample to capture random area effects. M-Quantile model determines an area effect by M-Quantile individual coefficients of the units belonging to the area. Assume that we have individual level data on y and x. Under linear M-quantile regression each sample value of (x, y) will lie on one and only one M-Quantile line. We refer to the q-value of this line as the M-Quantile coefficient of the corresponding sample unit. So, every sample unit will have an associate q-value. In order to estimate these unit-specific q-values, we define a fine grid of q-values (e.g., $0.001, \ldots, 0.999$) that adequately covers the conditional distribution of y and x. We fit an M-Quantile model for each q-value in the grid and use linear interpolation to estimate a unique q-value, q_j, for each individual j in the sample. This makes it possible to calculate an M-Quantile coefficient for each area by suitably averaging the q-values of each sampled individual in that areas. This area-specific q-value is denoted by $\hat{\theta}_d$ (Tzavidis et al. 2010).

Estimation of the MSE of estimator (2) can be achieved by using a linearization approach or a bootstrap approach proposed by Tzavidis et al. (2010). Other contributions on MSE estimation to be used with "pseudo-linear" small area estimators are those described in Chambers et al. (2011) and in Chambers et al. (2014).

Denoting by t the poverty line, the Foster, Greer, and Thorbecke (FGT) poverty measures for a small area d are defined as:

$$F_{\alpha d} = \frac{1}{N_d} \sum_{j=1}^{N_d} \left(\frac{t - y_{jd}}{t}\right)^{\alpha} I(y_{jd} \leq t). \tag{4}$$

The poverty line t is a level of income that defines the state of poverty (units with income below t are considered poor), y is a measure of income for individual/household j, N_d is the number of individuals/households in area d, I is the indicator function (equal to 1 when $y_{jd} \leq t$ and 0 otherwise), and α is a "sensitivity" parameter. When $\alpha = 0$, $F_{\alpha d}$ is the *HCR*, whereas when $\alpha = 1$, $F_{\alpha d}$ is the *Poverty Gap*.

Consider the following decomposition of the FGT indicators:

$$F_{\alpha d}^{MQ} = N_d^{-1} \left(\sum_{j \in s_d} F_{\alpha j d} + \sum_{j \in r_d} F_{\alpha j d} \right) \tag{5}$$

where $\sum_{j \in r_d} F_{\alpha j d}$ is unknown and needs to be predicted.

Referring to the decomposition (5), the question is how to estimate the out of sample component in that expression. This can be achieved using the same ideas described above for estimating the small area mean under the M-quantile small area model. Indeed, the estimation of poverty indicators is a special case of the quantile estimation since we are interested in estimating the number of individuals/households below a threshold. As a result, one approach to the estimation of the $F_{\alpha d}$ is the usage of a smearing-type estimator of the distribution function such as the CD estimator. In this case, an estimator $\hat{F}_{\alpha d}^{MQ}$ of $F_{\alpha d}^{MQ}$ is

$$\hat{F}_{\alpha d}^{\mathrm{MQ}} = N_d^{-1} \left\{ \sum_{j \in s_d} I(y_{jd} \leq t) + \sum_{k \in r_d} n_d^{-1} \sum_{j \in s_d} I(\hat{y}_{kd} + (y_{jd} - \hat{y}_{jd}) \leq t) \right\}, \quad (6)$$

that can be computed using a Monte Carlo procedure described in Pratesi et al. (2010), which is similar to that proposed by Molina and Rao (2010).

3   Poverty Mapping in Italy

In Italy, the European Survey on Income and Living Conditions (EU-SILC) is conducted yearly by ISTAT to produce estimates on the living conditions of the population at the national and regional (NUTS-2) levels (ISTAT 2008).

Regions are planned domains for which EU-SILC estimates are published, while the provinces are unplanned domains as well as municipalities that are partitions of the provinces. The regional samples are based on a stratified two stage sample design: in each province the municipalities are the primary sampling units (PSUs), while the households are the secondary sampling units (SSUs). The PSUs are divided into strata according to their dimension in terms of population size; the SSUs are selected by means of systematic sampling in each PSU. All the members of each sampled household are interviewed through an individual questionnaire, and one individual in each household (usually, the head of the household) is interviewed through a household questionnaire. It is useful to note that some provinces, generally the smaller ones, may have very few sampled municipalities; furthermore, many municipalities are not even included in the sample at all. Direct estimates may therefore have large errors at provincial level or they may not even be computable at municipality level, thereby requiring resort to SAE techniques.

Applying the small area methodologies presented in Sect. 2 to data from the EU-SILC 2008 survey requires covariate information that is also known for every not sampled household in the population. This information is available from the 2001 Population Census of Italy. We use these data under the hypothesis that the time lag of 7 years between EU-SILC and Census data does not play a determinant role on the estimates. Empirical results support this hypothesis (Pratesi et al. 2011). As an alternative, updated local administrative databases could be used. The Population Census of Italy has a very comprehensive questionnaire, collecting information on each household and on each individual living in the Italian territory. For the purpose of obtaining estimates on poverty and living conditions in Tuscany, we selected census variables that are also available from the EU-SILC survey. These variables were included as covariates in the working small area models used for estimating the mean household income and the poverty indicators. Thus, we can say that the EU-SILC datasets, together with data coming from the Population Census of Italy, represent a complete and valuable source of information that can be used for applying advanced SAE techniques for producing poverty and living condition estimates in Italy.[1]

More in details, in the working small area models employed for the present application, the equivalized household income is the outcome variable. Averages, percentiles, and poverty indicators are computed on the household equivalized disposable income.[2]

3.1 The Results in Three Italian Regions

Data on the equivalized income in 2007 are available from the EU-SILC survey 2008 for 1,495 households in the 10 Tuscany Provinces, for 1,286 households in the 5 Campania Provinces and for 2,274 households in the 11 Lombardia Provinces. A set of explanatory variables is available for each unit in the population from the Population Census 2001.

We chose the Campania, Lombardia, and Toscana regions because they are representative respectively of the South, Center, and North of Italy and they are able to illustrate the well-known North–South divide (Brandolini et al. 2007).

The boxplots (Fig. 1) show evidence of skewed distribution of the household equivalized income with heavy tail on the right in all the three regions. Evidence of outliers emerges from the boxplots and also from the summary statistics obtained using the cross-sectional EU-SILC household weights (see Table 1).

According to the Italian Census 2001 data collected by ISTAT, the Campania region accounts for 1,862,855 households, the Lombardia region accounts for 3,652,944 households, and the Toscana region accounts for 1,388,252 households. Among the available auxiliary variables relying on previous studies of poverty assessment we selected the following covariates to fit the small area models:

---

[1] As a remark, it is important to underline that EU-SILC and census data are confidential. These data were provided by ISTAT, the Italian National Institute of Statistics, to the researchers of the SAMPLE project and were analyzed for the present analysis respecting all confidentiality restrictions.

[2] This is computed as the Disposable Household Income multiplied by the Within-household non-response inflation factor and divided by the equivalized household size (EHS). The Disposable Household Income is the sum for all household members of gross personal income components *plus* gross income components at household level *minus* employer's social insurance contributions, interest paid on mortgage, regular taxes on wealth, regular inter-household cash transfer paid, tax on income, and social insurance contributions. The Within-household non-response inflation factor is the factor by which it is necessary to multiply the total gross income, the total disposable income, or the total disposable income before social transfers to compensate the non-response in individual questionnaires. The EHS is obtained as EHS $= 1 + 0.5 \cdot (\mathrm{HM}_{14+} - 1) + 0.3 \cdot \mathrm{HM}_{13-}$, where HM_{14+} is the number of household members aged 14 and over (at the end of income reference period), HM_{13-} be the number of household members aged 13 or less (at the end of income reference period). By this way we take into account the economy of scale present in an household (Eurostat 2007).

Fig. 1 Boxplots of the
disposable equivalized
household income

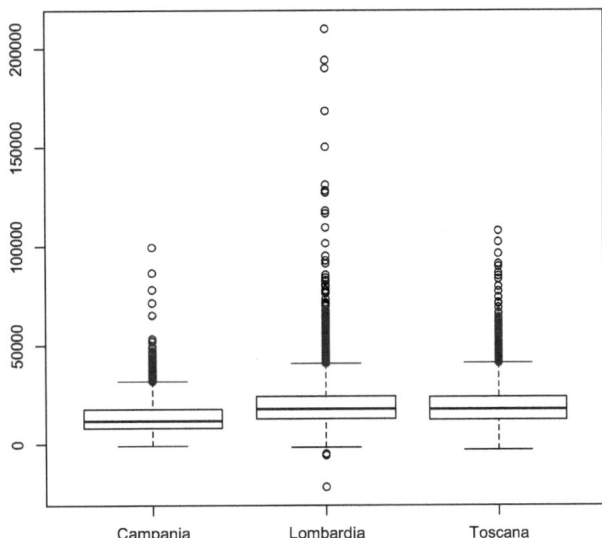

Table 1 Summary statistics for the household equivalized income

	Min.	1st Qu.	Median	Mean	3rd Qu.	Max.
Campania	−852.7	8,073	11,560	13,550	17,430	99,400
Lombardia	−21,550.0	12,620	17,670	20,040	24,000	209,800
Toscana	−2,849.0	12,120	17,230	19,430	23,570	107,900

household size (integer value), ownership of dwelling (owner/tenant), age of
the head of the household (integer value), years of education of the head of
the household (integer value), working position of the head of the household
(employed/unemployed in the previous week), and gender of the head of the
household.

We used an M-quantile model to estimate the HCR and Poverty Gap, the 20th
Percentile, the Median and the 80th Percentile, and the Mean at a LAU 1 level
(Provinces). The National poverty line is 9,310.74 Euros (equivalized household
income). It has been computed as the 60% of the median of the household
disposable equivalized income in Italy (21 regions). Here we report only a sample
of the entire set of the results: we focus on Poverty gap and HCR in the three Italian
regions using the national and regional poverty line. Additional results are available
and downloadable from the web site of the SAMPLE project, funded under the
7FP (http://www.sample-project.eu). An effective representation of the computed
poverty estimates is in Figs. 2 and 3.

The representation of the results by means of poverty mapping underlines the
importance of computing different poverty indicators to correctly depict the living
conditions in the areas of interest. In particular, it is interesting to accompany the
estimation of the HCR with that of the PG. Indeed, the HCR is useful to detect
the areas with higher percentages of poor households, while the PG helps to detect

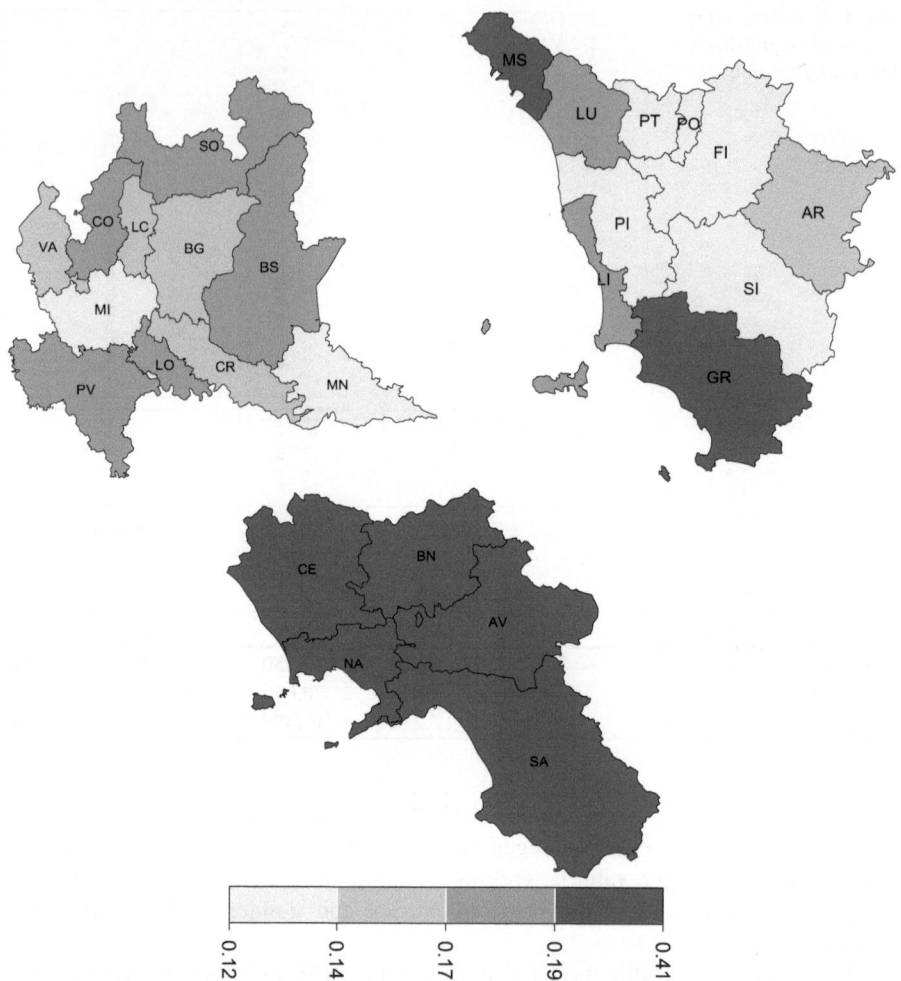

Fig. 2 Estimate of the head count ratio at provincial level Lombardia, Tuscany, and Campania

the areas where the intensity of poverty is lower. Thus, the PG gives an interesting indication to policy makers, since it suggests in which areas reducing the incidence of poverty, as measured by the HCR, would be easier. The results obtained under the national poverty line are a clear evidence of the North–South divide. The Campania Region appears to have the highest percentage of poors and the largest poverty gaps.

Tables 2, 3, and 4 show the estimates of the HCR, PG, and Median of the equalized income in the provinces of Lombardia, Toscana, and Campania region, respectively and using the national poverty line. The HCR in Lombardia ranges from 0.131 (Mantova) to 0.27 (Sondrio) while the PG for provinces in the same region ranges from 0.055 (Mantova) to 0.079 (Sondrio). For provinces in Toscana the HCR ranges from 0.221 (Massa-Carrara) to 0.124 (Prato) and the PG in the same region

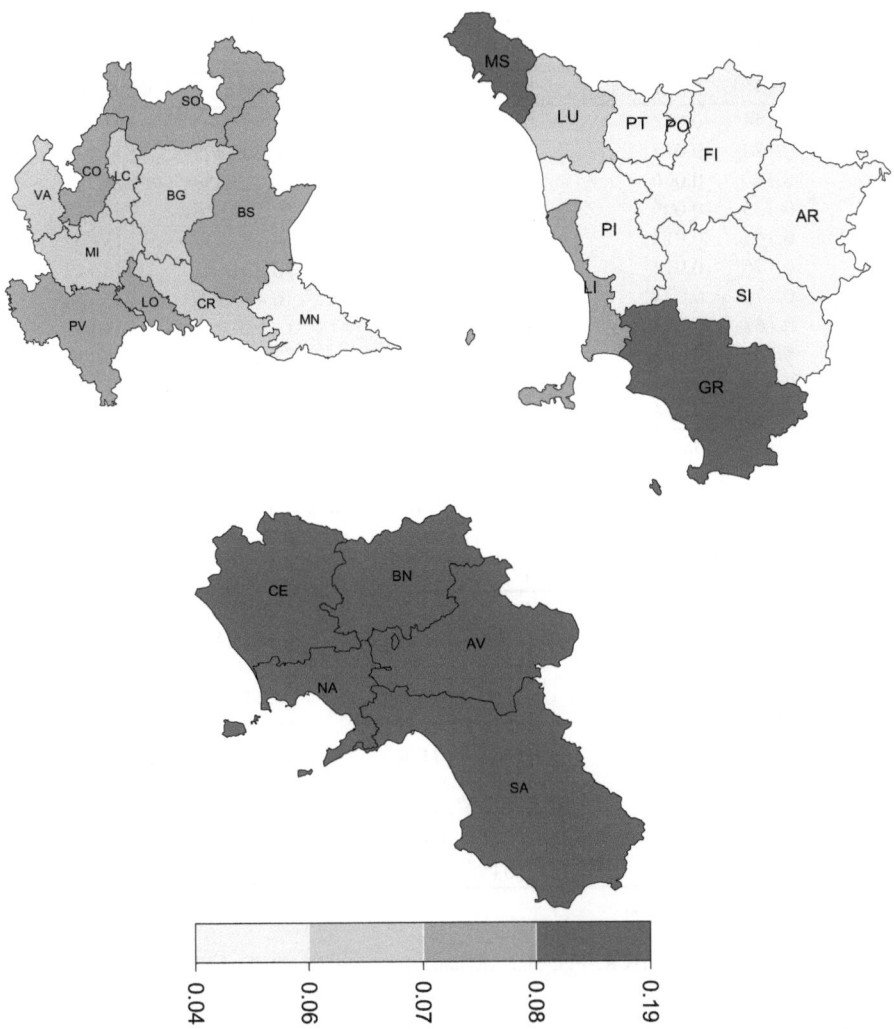

Fig. 3 Estimate of the poverty gap at provincial level Lombardia, Tuscany, and Campania

from 0.095 (Massa-Carrara) to 0.045 (Prato). Finally, for provinces in the region of Campania the HCR ranges from 0.335 (Salerno) to 0.411 (Caserta) and the PG from 0.138 (Salerno) to 0.189 (Caserta). Also in this case, the picture that emerges is as expected, i.e., Campania is a region that has consistently higher poverty than Toscana and Lombardia. The result is confirmed when taking into account the regional poverty lines for producing the estimates in Campania. It becomes apparent that the differences in poverty between the North of Italy and the South of Italy are even more pronounced (see the results of the SAMPLE project in Pratesi et al. 2010). The use of the PG indicator significantly enhances the picture of wealth in

Table 2 Estimates of the HCR, PG, and Median of the equalized income in the provinces of Lombardia region

Province	HCR	RMSE (HCR)	PG	RMSE (PG)	Median	RMSE (Median)
VA	0.144	0.011	0.060	0.006	17,797.703	433.110
CO	0.170	0.017	0.073	0.010	16,205.475	614.036
SO	0.180	0.036	0.079	0.021	16,719.356	1,429.371
MI	0.135	0.009	0.058	0.005	18,784.516	263.734
BG	0.162	0.013	0.069	0.008	16,759.831	515.092
BS	0.176	0.014	0.077	0.008	16,354.615	476.113
PV	0.170	0.024	0.073	0.013	16,459.965	1,037.750
CR	0.167	0.021	0.072	0.012	16,559.237	857.137
MN	0.131	0.018	0.055	0.010	17,833.713	585.117
LC	0.146	0.017	0.061	0.009	17,747.664	715.015
LO	0.176	0.025	0.077	0.014	16,281.596	1,013.570

Poverty line is 9,310.74 Euros

Table 3 Estimates of the HCR, PG, and Median of the equalized income in the provinces of Toscana region

Province	HCR	RMSE (HCR)	PG	RMSE (PG)	Median	RMSE (Median)
MS	0.221	0.027	0.095	0.016	14,530.520	771.250
LU	0.174	0.019	0.070	0.010	15,966.472	646.131
PT	0.143	0.017	0.054	0.009	17,365.758	663.371
FI	0.132	0.011	0.050	0.006	18,035.045	387.938
LI	0.185	0.021	0.076	0.012	15,921.874	753.767
PI	0.133	0.015	0.050	0.008	17,937.921	646.715
AR	0.146	0.016	0.056	0.009	16,937.315	651.878
SI	0.139	0.017	0.053	0.009	17,738.957	770.851
GR	0.194	0.026	0.081	0.015	16,102.594	967.963
PO	0.124	0.017	0.045	0.009	18,216.393	713.084

Poverty line is 9,310.74 Euros

Table 4 Estimates of the HCR, PG, and Median of the equalized income in the provinces of Campania region

Province	HCR	RMSE (HCR)	PG	RMSE (PG)	Median	RMSE (Median)
CE	0.411	0.024	0.189	0.017	10,430.029	538.140
BN	0.387	0.041	0.163	0.026	10,428.849	860.884
NA	0.384	0.011	0.177	0.008	11,454.023	217.499
AV	0.349	0.027	0.144	0.018	11,675.029	619.829
SA	0.335	0.020	0.138	0.013	12,384.882	444.130

Poverty line is 9,310.74 Euros

the different regions. Noticeable are also some aspects of the comparison between the regions of Toscana and Lombardia. Previous analyses indicated that Lombardia is somewhat wealthier than Toscana. However, looking at the estimates of HCR and PG a different picture emerges. Not including Massa-Carrara and Grosseto, provinces in Toscana have lower HCR and PG than provinces in Lombardia. For example, Pavia, one of the wealthiest provinces, in terms of average income and income distribution, in Lombardia appears to have higher poverty than a number of provinces in Toscana such as Siena and Florence.

Of course, in these comparisons one must take into account the precision of the estimates and the fact that each region has its own poverty line that is different from the national poverty line. Nevertheless, these results indicate that inequalities in Lombardia may be more pronounced than inequalities in Toscana.

Mean Squared Error estimation for the parameter of interest (HCR, PG, and Median) is discussed in detail in the paper by Marchetti et al. (2012) and is based on a nonparametric bootstrap scheme. For short, starting from sample s, selected from a finite population Ω without replacement, we fit the M-quantile small area model from which we obtain the model residuals. We then generate B bootstrap populations, Ω^b. From each bootstrap population we select L bootstrap samples, using simple random sampling within the small areas and without replacement. Using the bootstrap samples we obtain estimates of the target poverty indicators. Bootstrap populations are generated by sampling from the empirical distribution of the residuals, or a smoothed version of this distribution, conditionally or unconditionally on the small areas. Bootstrap estimators of the bias and variance of the estimated target small area parameter, $\hat{\tau}_j$, are defined respectively by:

$$\hat{B}(\hat{\tau}_j) = B^{-1}L^{-1}\sum_{b=1}^{B}\sum_{l=1}^{L}(\hat{\tau}_j^{*bl} - \tau_j^{*b}) \qquad \hat{V}(\hat{\tau}_j) = B^{-1}L^{-1}\sum_{b=1}^{B}\sum_{l=1}^{L}(\hat{\tau}_j^{*bl} - \bar{\hat{\tau}}_j^{*bl})^2,$$

where τ_j^{*b} is the small area parameter of the b-th bootstrap population, $\hat{\tau}_j^{*bl}$ is the small area parameter estimated by using the l-th sample from the b-th bootstrap population, and $\bar{\hat{\tau}}_j^{*bl} = L^{-1}\sum_{l=1}^{L}\hat{\tau}_j^{*bl}$. The bootstrap MSE estimator of the estimated small area target parameter is then defined as:

$$\hat{M}(\hat{\tau}_j) = \hat{V}(\hat{\tau}_j) + \hat{B}(\hat{\tau}_j)^2.$$

In the application we set $B = 50$ and $L = 100$ and we used the empirical unconditional approach.

4 Extensions of M-Quantile Regression Model and Final Remarks

The methodology we propose improves the traditional poverty mapping methods in the following ways: (a) it enables the estimation of the distribution function and percentiles of the study variable within the small area of interest, (b) it provides analytical, and also empirical, estimation of the mean squared error of the M-quantile small area mean estimates, and (c) it employs a robust to outliers estimation method.

Detailed information at local level can be used efficiently by the policy makers to develop "ad-hoc" interventions against poverty. This is an important goal that can be accomplished more easily when the distribution of income and the inequalities are appreciated at local level.

Until very recently the practice of poverty mapping has been driven by the World Bank method proposed by Elbers et al. (2003). It combines individual/household survey data and population data with the objective of estimating poverty indicators for specific geographic areas as small as villages or hamlets. More recently, researchers in the SAE field have intensively studied the World Bank method and have proposed small area models for poverty mapping. A very recent method is the empirical best prediction (EBP) approach proposed by Molina and Rao (2010). In a previous paper (Pratesi et al. 2013) we focused on limits and potentialities of the estimators obtained under this approach in comparison with M-quantile method. Here we focused on the characteristics of the M-quantile approach to SAE.

Indeed the MQ/CD estimators of poverty indicators presented here are in some sense automatically robust against outlier observations. However, robust EBLUP by Sinha and Rao (2009) and robust M-quantile estimators by Giusti et al. (2011) are recent proposals to consider the role played by outlying values. The M-quantile approach to poverty estimation does not require strong distributional assumptions and, because it is in some sense automatically robust to the presence of outlying observations, it can be applied directly to raw income data.

The absence of distributional assumption makes it easier to extend the approach to nonparametric modeling and to the inclusion of spatial information in poverty mapping.

M-quantile models do not depend on strong distributional assumptions, but they assume that the quantiles of the distribution are some known parametric function of the covariates. When the functional form of the relationship between the q-th M-quantile and the covariates deviates from the assumed one, the traditional M-quantile regression can lead to biased estimators of the small area parameters. Pratesi et al. (2008a,b) and Salvati et al. (2011) have extended this approach to the M-quantile method for the estimation of the small area parameters using a nonparametric specification of the conditional M-quantile of the response variable given the covariates. Using penalized splines (p-splines) for M-quantile regression, beyond having the properties of M-quantile models, allows for dealing with an undefined functional relationship that can be estimated from the data. When the relationship between the q-th M-quantile and the covariates is not linear, a p-splines M-quantile regression model may have significant advantages compared to the linear M-quantile model (see also Pratesi et al. 2008a,b).

Typically, random effects models assume independence of the random area effects. This independence assumption is also implicit in M-quantile small area models. In economic applications, however, observations that are spatially close may be more related than observations that are further apart. This spatial correlation can be accounted for by extending the random effects model to allow for spatially correlated area effects using, for example, a simultaneous autoregressive (SAR) model (see Petrucci and Salvati 2006; Pratesi and Salvati 2008, 2009).

An alternative approach to incorporate the spatial information in the regression model is by assuming that the regression coefficients vary spatially across the geography of interest. GWR (see Brunsdon et al. 1996) extends the traditional regression model by allowing local rather than global parameters to be estimated. Salvati et al. (2012) proposed an M-quantile GWR small area model. The extension provides a locally robust model for the M-quantiles of the conditional distribution of the outcome variable given the covariates.

Design-based properties of M-quantile estimators are not completely explored. The M-quantile approach does not allow for the use of unit level survey weights, making questionable the design consistency of the estimators unless the sampling design is self-weighting within small areas (Fabrizi et al. 2014). Another issue is benchmarking, that is the consistency of a collection of small area estimates with a reliable estimate obtained according to ordinary design-based methods for the union of the areas (Fabrizi et al. 2012).

It is important to accompany the small area estimates with a measure of their variability. When we refer to empirical estimation of variance, we note that bootstrap approach to the estimate of MSE is faster than bootstrap for EBLUP (mixed linear model require double bootstrap techniques) (Marchetti et al. 2012). Also analytical estimators of MSE have been developed. For each point estimator (HCR, PG, and quantiles of the income distribution) research has been done to present an estimator for the corresponding root mean squared error. In the economy of this work we decided not to present these estimators in details. Useful references are in Pratesi et al. (2010) and also in successive works by Chambers et al. (2011, 2014).

Finally, further interesting developments and applications of the M-quantile approach will regard the computation at the small area level of non-monetary measures of poverty, such as fuzzy poverty indicators to get a more complete picture of the poverty situation in the areas of interest (Cheli and Lemmi 1995). At the moment there is no specification of M-quantile model if the response variable is multivariate or discrete. Further research will then focus on methods both for binary and count outcomes and will further develop and evaluate approaches to estimating the precision of the domain estimates derived under these new methodologies.

Acknowledgements This work was financially supported by the European Project SAMPLE "Small Area Methods for Poverty and Living Condition Estimates," European Commission 7th FP—www.sample-project.eu. My personal thanks to Caterina Giusti, Stefano Marchetti, and Nicola Salvati for their assistance and contributions.

References

Brandolini, A., Saraceno, C.: Introduzione. In: Brandolini, A., Saraceno, C. (eds.) Povertá e Benessere. Una geografia delle disuguaglianze in Italia. Il Mulino, Bologna (2007)
Brunsdon, C., Fotheringham, A.S., Charlton, M.: Geographically weighted regression: a method for exploring spatial nonstationarity. Geographical Anal. **28**, 281–298 (1996)

Chambers, R.L., Dunstan, R.: Estimating distribution functions from survey data. Biometrika **73**, 597–604 (1996)

Chambers, R.L., Pratesi, M.: Small area methodology in poverty mapping: An introductory overview. In: Lemmi, A., Betti, G. (eds.) Poverty and Social Exclusion: New Methods of Analysis. Routledge, London (2013)

Chambers, R.L., Tzavidis, N.: M-quantile models for small area estimation. Biometrika **93**, 255–268 (2006)

Chambers, R.L., Chandra, H., Tzavidis, N.: On bias robust mean squared error estimation for pseudolinear small area estimators. Surv. Methodol. **37**(2), 153–170 (2011)

Chambers, R.L., Chandra, H., Salvati, N., Tzavidis, N.: Outliers robust small area estimation. J. R. Stat. Soc. B **76**, 47–69 Part 1 (2014)

Cheli, B., Lemmi, A.: A totally fuzzy and relative approach to the multidimensional analysis of poverty. Econ. Notes **24**, 115–134 (1995)

Elbers, C., Lanjouw, J.O., Lanjouw, P.: Micro-level estimation of poverty and inequality. Econometrica **71**, 355–364 (2003)

Eurostat: Description of SILC user database variables. Version 2007.1 from 01-03-2009. Eurostat (2007)

Fabrizi, E., Salvati, N., Pratesi, M.: Constrained small area estimators based on M-quantile methods. J. Off. Stat. **28**(1), 89–106 (2012)

Fabrizi, E., Salvati, N., Pratesi, M., Tzavidis, N.: Outlier robust model-assisted small area estimation. Biom. J. **56**(1), 157–175 (2014)

Foster, J., Greer, J., Thorbecke, E.: A class of decomposable poverty measures. Econometrica **52**, 761–766 (1984)

Giusti, C., Pratesi, M., Salvati, N.: Small area methods in the estimation of poverty indicators: The case of Tuscany. Polit. Econ. **3**, 369–380 (2009)

Giusti, C., Tzavidis, N., Pratesi, M., Salvati, N.: Resistance to outliers of M-quantile and robust random effects small area models. S3RI Methodology Working Papers **M11/04**. Southampton Statistical Sciences Research Institute, University of Southampton (2011)

ISTAT: L'indagine europea sui redditi e le condizioni di vita delle famiglie (Eu-Silc). Metodi e Norme **37**, 1–191 (2008)

Marchetti, S., Tzavidis, N., Pratesi, M.: Nonparametric bootstrap mean squared error estimation for M-quantile estimators for small area averages, quantiles and poverty indicators. Comput. Stat. Data Anal. **56**, 2889–2902 (2012)

Molina, I., Rao, J.N.K.: Small area estimation of poverty indicators. Can. J. Stat. **38**, 369–385 (2010)

Petrucci, A., Salvati, N.: Small area estimation for spatial correlation in watershed erosion assessment. J. Agric. Biol. Environ. Stat. **11**, 169–182 (2006)

Pratesi, M., Salvati, N.: Small area estimation: the EBLUP estimator based on spatially correlated random area effects. Stat. Methods Appl. **17**, 113–141 (2008)

Pratesi, M., Salvati, N.: Small area estimation in the presence of correlated random area effects. J. Off. Stat. **25**, 37–53 (2009)

Pratesi, M., Ranalli, M.G., Salvati, N.: Semiparametric M-quantile regression using penalized splines. J. Nonparametr. Stat. **21**, 287–304 (2008a)

Pratesi, M., Ranalli, M.G., Salvati, N.: Semiparametric M-quantile regression for estimating the proportion of acidic lakes in 8-digit HUCs of the North-eastern US. Environmetrics **19**, 687–701 (2008b)

Pratesi, M., Giusti, C., Salvati, N.: Estimation of poverty indicators in Italy: A small area M-quantile based approach. Riv. Ital. Econ. Demogr. Stat. **LXIII**(1–2), 243–256 (2009)

Pratesi, M., Giusti, C., Marchetti, S., Salvati, N., Tzavidis, N., Molina, I., Durbán, M., Grané, A., Marín, J.M., Veiga, M.H., Morales, D., Esteban, M.D., Sánchez, A., Santamaría, L., Marhuenda, Y., Pérez, A., Pagliarella, M.C., Rao, J.N.K., Ferretti, C.: Sample deliverables 12 and 16. Final small area estimation developments and simulations results. Available at http://www.sample-project.eu/it/the-project/deliverables-docs.html (2010)

Pratesi, M., Giusti, C., Marchetti, S., Salvati, N.: Robust small area estimation of poverty indicators. In: Survey Research Methods and Applications. Edizioni Plus, Pisa (2011)

Pratesi, M., Marchetti, S., Giusti, C.: Small area estimation of poverty indicators. In: Davino, C., Fabbris, L. (eds.) Survey Data Collection and Integration. Springer, New York (2013)

Rao, J.N.K.: Small Area Estimation. Wiley, New York (2003)

Salvati, N., Ranalli, M.G., Pratesi, M.: Small area estimation of the mean using nonparametric M-quantile regression: A comparison when a linear mixed model does not hold. J. Stat. Comput. Simul. **81**(8), 945–964 (2011)

Salvati, N., Tzavidis, N., Pratesi, M., Chambers, R.: Small area estimation via M-quantile geographically weighted regression. TEST **21**(1), 1–28 (2012)

Salvati, N., Giusti, C., Pratesi, M.: The use of spatial information for the estimation of poverty indicators at the small area level. In: Lemmi, Betti (eds.) Poverty and Social Exclusion: New Methods of Analysis. Routledge, London (2013)

Sinha, S.K., Rao, J.N.K.: Robust small area estimation. Can. J. Stat. **37**, 381–399 (2009)

Tzavidis, N., Salvati, N., Pratesi, M., Chambers, R.: M-quantile models with application to poverty mapping. Stat. Methods Appl. **17**(3), 393–411 (2008)

Tzavidis, N., Marchetti, S., Chambers, R.: Robust prediction of small area means and distributions. Aust. N. Z. J. Stat. **52**, 167–186 (2010)

Use of Zero Functions for Combining Information from Multiple Frames

A.C. Singh and F. Mecatti

Abstract A class of generalized multiplicity-adjusted Horvitz–Thompson (GMHT) estimators was introduced by Singh and Mecatti (J. Official Stat. 27(4):633–650, 2011, JOS) to provide a unified and systematic approach to existing estimators via GMHT-Regression. The main purpose of this chapter is to present key observations that led to the development of the unified principled framework. These are based on the use of zero functions as predictors in regression which play a fundamental role in statistical estimation such as quasi-likelihood. The key observations are listed below. First, an optimal combination of two estimators of the same domain is equivalent to regression of the simple multiplicity-adjusted HT (i.e., average of the two estimators) on the zero function defined by the difference of two estimators. Second, there are two types of zero functions for each overlapping domain—one based on domain count estimates and the other based on domain total estimates. Some estimators use both types of zero functions as predictors in regression while others optimally combine first the two domain count estimates and then apply Hájek-type ratio adjustments for each sample to optimally estimated domain counts before regressing on domain total zero functions. Third, the regression need not be optimal as it can be based on a working covariance structure to obtain robust consistent estimates; this is in view of the observation that optimal regression estimators are typically unstable (i.e., with high coefficient of variation) for complex designs due to lack of adequate degrees of freedom for estimating regression parameters. Fourth, for any regression estimator, it may be better to use count zero functions via Hájek-adjustment first because the initial

A.C. Singh (✉)
Center for Excellence in Survey Research, NORC at the University of Chicago, Chicago,
IL 60603, USA
e-mail: singh-avi@norc.org

F. Mecatti
Dipartimento di Sociology and Social Research, Universita di Milano-Bicocca,
20216 Milano, Italy

F. Mecatti et al. (eds.), *Contributions to Sampling Statistics*, Contributions to Statistics,
DOI 10.1007/978-3-319-05320-2_3,
© Springer International Publishing Switzerland 2014

GMHT estimator may not be well correlated with count zero functions. Fifth, it might be preferable to use a suitable working covariance matrix than the optimal one in order to obtain a calibration form in addition to making the estimator more stable. Finally, sixth, the basic principles used in the unified framework make it possible to construct new improved estimators over other estimators in the literature. GMHT-Regression estimators are constructed as sums of contributions from each frame which allow for application of standard variance estimation techniques.

Keywords Hájek-ratio adjustment • Multiplicity adjustment • Optimal regression • Random controls • Working covariance matrix

1 Introduction

Before considering multiple frames (MF), we review the problem of regression estimation in survey sampling for single frames (SF) and consider how predictor zero functions play a key role in improving precision of basic Horvitz–Thompson (HT) estimators. We assume for simplicity that the initial HT estimator is already adjusted for nonresponse. Also we assume initially that there is no coverage bias and the goal of estimation is to reduce variance although in practice the two go hand-in-hand. For SF, regression estimation transforms the basic HT estimator to an HT-Regression estimator such as generalized regression (GREG)—a popular commonly used method in survey sampling; see, e.g., Särndal (1980) and Fuller (1975). After reviewing regression estimators for SF, we review MF estimators and consider how a simple multiplicity-adjusted HT (SMHT) estimator can be transformed via regression on old predictor zero functions (these are traditional for SF) and new ones (nontraditional for MF from overlap frame domains) to produce a class of SMHT-Regression estimators which includes various existing estimators. In particular, we emphasize GROUM(s) and GROUM(g) (Singh and Wu 2003 although termed differently here) which are the MF analogues of the recently developed SF calibration estimator termed generalized raking with optimal unbiased modification (GROUM) by Singh et al. (2013). GROUM is a generalization of GREG to build in optimality and to produce range-restricted weight adjustment factors; the extension "s" or "g" in the MF version denote use of either simple or general multiplicity adjustment factors (*mafs*) in GROUM.

The main purpose of this chapter is to review basic principles of regression estimation in survey sampling using predictor zero functions for both SF and MF and then consider the calibration approach to adjust for coverage bias in MF estimators analogous to SF calibration estimators. The organization of the chapter is as follows. Section 2 provides a review of regression on zero functions for combining information from SF which includes survey data from two time points for partially overlapping panel surveys. For commonly used stratified designs, we also review the recently developed estimator GROUM to build variance optimality in regression calibration as an alternative to traditional optimal regression estimators which tend to be unstable and do not lend themselves in general to a calibration form. To this end, relative design adjustment factors (*dafs*) are introduced in the regression working covariance structure to capture possibly different designs in different strata and are specified outside the calibration estimating equations by minimizing the resulting unequal weighting effect (UWE) after calibration via a grid search. In Sect. 3, we consider the problem of two frames with partial overlap and review GROUM(*s*) and GROUM(*g*) as MF versions of GROUM for SF. For building optimality in GROUM(*s*) or (*g*), we introduce relative *dafs* to account for possibly different designs in different frames in MF analogous to different strata in SF. The calibration approach to GROUM(*s*) and GROUM(*g*) via a coverage bias model is considered in Sect. 4. Section 5 contains generalizations to three or more frames along with concluding remarks.

2 Review of Zero Functions and GREG-Type Estimation for Single Frames

The term zero function (or an elementary estimating function) is used to denote a function of data or parameters or both that is an unbiased estimate of zero. The theory of estimating functions, also known as quasi-likelihood (QL), provides a very powerful semi-parametric framework (i.e., under only first two moment assumptions) for producing optimal estimators for linear and nonlinear models with or without random effects; see McCullagh and Nelder (1989, Chap. 9), Godambe and Thompson (1989, 2009, Chap. 26) and also Singh and Rao (1997) for a simple review and some generalizations. For complex surveys, pseudo maximum likelihood (Binder 1983; Skinner 1989, p. 80) or weighted QL (Singh 2009, Chap. 35) are typically used for super-population parameters. For estimating finite population parameters, often a working assumption is made about second moments of HT-type estimating functions due to lack of second order inclusion probabilities or instability of the estimated covariance matrix (in the sense of high coefficient of variation for each element of the matrix), and although theoretically the resulting estimators become sub-optimal, they do possess good finite sample properties. Moreover, they encompass commonly used estimators such as GREG well known for wide applications. In the following, we provide more details of the above principled strategy of estimation for simple and complex surveys.

2.1 Super-Population Parameters

First consider simple survey designs; i.e., simple random sampling (SRS) without replacement with negligible sampling fractions. If the model is linear with fixed parameters β, then the Gauss–Markov theorem provides best linear unbiased estimates $\widehat{\beta}_{\text{BLUE}}$ and the optimal estimating equation is given by

$$X'\Sigma^{-1}(y - X\beta) = 0, \tag{1}$$

where all the terms are defined in the standard way. Let n denote the number of observations and p the number of $\boldsymbol{\beta}$-parameters. The estimating Eq. (1) provides an optimal combination of n zero functions $\boldsymbol{y} - \boldsymbol{X}\boldsymbol{\beta}$ that provide a set of p zero functions to estimate the p-dimensional $\boldsymbol{\beta}$-parameters. Equivalently, we can sacrifice the first p observations to obtain an initial estimator $\widehat{\boldsymbol{\beta}}^{(0)}$, and then regress the zero function $\boldsymbol{\beta} - \widehat{\boldsymbol{\beta}}^{(0)}$ on the n–p remaining zero functions, $\left(y_i - \boldsymbol{x}_i'\widehat{\boldsymbol{\beta}}^{(0)}\right), p+1 \leq i \leq n$ to obtain the final estimator $\widehat{\boldsymbol{\beta}}_{\text{BLUE}}$. Note that the combination of zero functions in the estimating function (1) becomes exactly zero at the estimator $\widehat{\boldsymbol{\beta}}_{\text{BLUE}}$ because the estimator solves the above estimating equation.

For semi-parametric nonlinear models, the optimal estimating function method or the QL method provides $\widehat{\boldsymbol{\theta}}_{\text{QL}}$ as the unique solution of the following equation under usual regularity conditions:

$$-E\left(\frac{\partial \boldsymbol{\mu}\left(\boldsymbol{\theta}\right)}{\partial \boldsymbol{\theta}'}\right)' \boldsymbol{\Sigma}(\boldsymbol{\theta})^{-1} \left(\boldsymbol{y} - \boldsymbol{\mu}\left(\boldsymbol{\theta}\right)\right) = \boldsymbol{0}, \tag{2}$$

where the covariance matrix $\boldsymbol{\Sigma}(\boldsymbol{\theta})$ may depend on $\boldsymbol{\theta}$-parameters as in the case of count data under log linear models. The BLUE estimating Eq. (1) is a special case of QL-estimating Eq. (2) when the model is linear and the covariance matrix $\boldsymbol{\Sigma}(\boldsymbol{\theta})$ does not depend on $\boldsymbol{\theta}$. In the case of parametric models, the score function is an optimal estimating function which yields maximum likelihood estimates. It may be noted that even with the matrix $\boldsymbol{\Sigma}(\boldsymbol{\theta})$ being a working covariance matrix, the estimating Eq. (2) provides consistent estimates of fixed $\boldsymbol{\theta}$-parameters as long as the model for the mean is correct; the resulting estimating functions are of course non-optimal.

For complex surveys, the weighted QL is used where the census estimating function obtained from (2) based on the whole data from the finite population of size N is estimated by suitably weighted sample estimating functions. The corresponding estimates possess certain optimality properties (Godambe and Thompson 1986).

2.2   Finite Population Parameters and GREG Estimators

For estimating the finite population total parameter T_y, instead of working with unit level data $\boldsymbol{y}$ and auxiliary variables $\boldsymbol{x}$, it is preferred to work with aggregate level summary statistics (Horvitz–Thompson estimators) $t_{y,\text{HT}}$ and $t_{x,\text{HT}}$ for design-based inference. The summary HT-estimators provide unbiased elementary estimating or zero functions $t_{y,\text{HT}} - T_y$, and $t_{x,\text{HT}} - \boldsymbol{T}_x$ with mean 0 under the complex design— a critical requirement for design-consistent estimation. The auxiliary totals $\boldsymbol{T}_x$ are assumed to be known. Denoting by $\boldsymbol{V}$ the estimated design-based covariance matrix of the vector, $(t_{y,\text{HT}} - T_y, (t_{x,\text{HT}} - \boldsymbol{T}_x)')'$, the QL estimating equation using (2) is given by

$$\left(1 \; \boldsymbol{0}'\right) \boldsymbol{V}^{-1} \begin{pmatrix} t_{y,\text{HT}} - T_y \\ t_{x,\text{HT}} - \boldsymbol{T}_x \end{pmatrix} = \boldsymbol{0}. \tag{3}$$

The resulting estimator is the optimal regression (OR) estimator:

$$t_{y,\mathrm{OR}} = t_{y,\mathrm{HT}} - \widehat{\boldsymbol{\beta}}'_{\mathrm{OR}} \left(t_{x,\mathrm{HT}} - \boldsymbol{T}_x \right). \tag{4}$$

The estimated covariance matrix $\boldsymbol{V}$ is generally unstable due to lack of sufficient degrees of freedom under complex designs. In practice, it is often the case that a suitable working covariance matrix $\tilde{\boldsymbol{V}}$ leads to favorable performance of the estimator. As shown in Singh (1996), if $\tilde{\boldsymbol{V}}$ is chosen as the SRS-type covariance matrix without centering but with sampling weights, then Eq. (4) gives rise to the well-known GREG (generalized regression) estimator which generalizes the regression methodology to complex designs. However, to make the working covariance matrix more suitable by centering, the initial sampling weights are Hájek-ratio adjusted to the known population count N so that they sum to N as in SRS. As an example, consider the working value of the estimated covariance between $t_{y,\mathrm{HT}}$ and the ith x-variable total estimator $t_{xi,\mathrm{HT}}$ as

$$\mathrm{cov}\left(t_{y,\mathrm{HT}}, t_{xi,\mathrm{HT}}\right) = \left(N/\left(n-1\right)\right)\left(1 - n/N\right) \sum_{k=1}^{n} y_k \left(x_{ik} - \overline{x}_{iw}\right) w_k, \tag{5}$$

where w_k's are sampling weights that sum to N and $\overline{x}_{iw}$ is the sample weighted average of x_i. Observe that for SRS, w_k equals N/n and the expression in (5) reduces to the usual SRS covariance estimate. With the above type of working covariance and variance defined for all elements of $\tilde{\boldsymbol{V}}$ when all estimators are Hájek-ratio adjusted, the estimating Eq. (3) gives rise to the GREG estimator given below if the unit vector is one of the x-variables; i.e., the population count is in the vector $\boldsymbol{T}_x$ of control totals.

$$t_{y,\mathrm{GREG}} = t_{y,\mathrm{HT}} - \widehat{\boldsymbol{\beta}}'_{\mathrm{GREG}} \left(t_{x,\mathrm{HT}} - \boldsymbol{T}_x \right), \tag{6}$$

where $\widehat{\boldsymbol{\beta}}_{\mathrm{GREG}} = \left(\boldsymbol{X}'\boldsymbol{W}\boldsymbol{X}\right)^{-1}\boldsymbol{X}'\boldsymbol{W}\boldsymbol{y}$, $\boldsymbol{W}$ is the diagonal matrix of sampling weights, and the element of $t_{x,\mathrm{HT}}$ corresponding to the x-variable with all values being unity is simply $\widehat{N} = \sum_{k=1}^{n} w_k$. Interestingly for GREG, it is not necessary to assume that the initial weights are Hájek-adjusted to N for centering as long as the unit vector is one of the x-variables. Clearly, $t_{y,\mathrm{GREG}}$ has a calibration form in that $t_{y,\mathrm{GREG}} = \sum_{k=1}^{n} y_k w_{k,\mathrm{GREG}}$ where

$$w_{k,\mathrm{GREG}} = w_k \left(1 + \boldsymbol{x}'_k \widehat{\boldsymbol{\lambda}}_{\mathrm{GREG}}\right), \widehat{\boldsymbol{\lambda}}_{\mathrm{GREG}} = \left(\boldsymbol{X}'\boldsymbol{W}\boldsymbol{X}\right)^{-1} \left(\boldsymbol{T}_x - t_{x,\mathrm{HT}}\right). \tag{7}$$

Originally, the GREG estimator was derived as a model-assisted estimator under a super-population model $y_k = \boldsymbol{x}'_k \boldsymbol{\beta} + \varepsilon_k$, ε_ks being independent with mean 0 and common variance. The $\boldsymbol{\beta}$-parameters are first estimated using weighted QL to obtain $\widehat{\boldsymbol{\beta}}_{\mathrm{GREG}}$, and then the prediction estimator $\boldsymbol{T}'_x \widehat{\boldsymbol{\beta}}_{\mathrm{GREG}}$ is adjusted for design bias using the unbiased HT estimator $\sum_{k=1}^{n} \widehat{\varepsilon}_k w_k$ where $\widehat{\varepsilon}_k = y_k - \mathbf{x}'_k \widehat{\boldsymbol{\beta}}_{\text{GREG}}$. Clearly this is the same as $t_{y,\text{GREG}}$ of (6). However, the model-assisted formulation of GREG is not conducive for its generalization to two time points for composite estimation (see next section) or to two frames considered in this chapter. Hence we prefer to use the QL estimating function framework based on HT-estimators as summary statistics and a working covariance matrix for complex surveys.

2.3 Regression Composite Estimation for Data from Two Time Points

Using the alternative formulation of GREG in terms of QL estimation with summary statistics and a working covariance matrix mentioned above, it is possible to define GREG-type estimators for regression composite estimation. Here data from two successive time points $\tau - 1$ and τ under a partially overlapping panel survey are combined to obtain composite estimates with more efficiency than those based on only the cross-sectional data. The new predictor zero functions are based on matched subsamples from the current τ and the previous time point $\tau - 1$. From data on each key outcome (y) variable from the two time points, two new zero functions termed level-driven and change-driven are created. The level-driven predictor zero function is defined as the difference between two estimates of the total of the previous time point variable $y_{\tau-1}$ but defined retrospectively for the current time population total parameter to be denoted by $T_{y(\tau-1)|\tau}$. One is based on the matched subsample of the current sample for which $y_{\tau-1}$ values are assumed to be available either exactly after micro-matching or after imputation if the individual was nonrespondent at $\tau - 1$ or there is a new addition to the current population. Other is based on the larger full sample already observed estimator from $\tau - 1$ except that the sample is recalibrated to current geo-demographic population controls. The change-driven predictor zero function is also defined as the difference between two estimates of $T_{y(\tau-1)|\tau}$. Here, one is based on the current full sample estimator of $T_{y(\tau)}$ adjusted by adding the difference of an estimate of $T_{y(\tau-1)|\tau}$ based on the matched subsample of the current sample and an estimate of $T_{y(\tau)}$ also based on the matched subsample of the current sample, while other is simply the full sample estimator of $T_{y(\tau-1)|\tau}$ which is the same as in the level-driven predictor.

It follows that the number of new predictors can be large—twice as many as the number of key outcome variables selected as important covariates for any arbitrary study variable. The theory of modified regression (MR) of Singh (1996) was used to combine new predictor zero functions with old predictors used in GREG under a QL estimation framework with a working covariance matrix in which the random controls corresponding to full sample estimates of $T_{y(\tau-1)|\tau}$ were treated as nonrandom for the purpose of point estimation along with a GREG-type working covariance matrix. However, for variance estimation, extra variability due to random controls was accounted for. Above was a major application of the regression composite estimation methodology to the Canadian Labour Force

Survey; see Singh et al. (2001) for more details. To reduce the number of new zero functions (which are double the number of key outcome variables selected), and to bring more stability in the composite estimates, Fuller and Rao (2001) suggested an appealing version of MR in which pre-specified convex linear combinations of level- and change-driven predictors were used. This was the version that was eventually implemented for the Canadian Labour Force Survey in January 2000.

2.4 Modified GREG for Variance Optimality

Unlike the OR estimator (4), the GREG formulation has no built-in variance optimality although it would be desirable in some capacity because our goal here is to increase precision using auxiliary variables. On the other hand, GREG has the desirable calibration form which OR does not have in general; see Rao (1994). It follows from the alternative GREG formulation of Sect. 2.2 based on the SRS-type working covariance that GREG is optimal for SRS if the unit vector is one of the covariates. However, it is not optimal even for stratified SRS (STSRS) as the working covariance matrix does not recognize possible differences in designs for different strata, in particular even different sample sizes. To overcome this problem for stratified designs which are in common use in practice, Singh, Ganesh and Lin (2013) proposed to introduce stratum-specific relative design adjustment factor (*daf*) ζ_h, $0 < \zeta_h < 1$ for each stratum h such that $\sum_{h=1}^{H} \zeta_h = 1$. The working value for the covariance in (5) for the new regression estimator termed GROUM is defined up to a proportionality constant as

$$\mathrm{cov}\left(t_{y,\mathrm{HT}}, t_{xi,\mathrm{HT}}\right) \propto \sum_{h=1}^{H} \zeta_h^{-1} \sum_{k=1}^{n_h} y_{hk} \left(x_{ihk} - \overline{x}_{ihw}\right) w_{hk}, \qquad (8)$$

where n_h is the stratum h sample size, w_{hk}'s are stratum-specific sampling weights, and other quantities are defined in the usual manner. With the above choice of the working covariance and provided each stratum indicator vector with 1s for units in the stratum and 0s elsewhere are part of the covariates with corresponding known population counts N_h, the GROUM estimator, analogous to GREG, is given by

$$t_{y,\mathrm{GROUM}} = t_{y,\mathrm{HT}} - \widehat{\boldsymbol{\beta}}'_{\mathrm{GROUM}} \left(t_{x,\mathrm{HT}} - \boldsymbol{T}_x\right), \qquad (9)$$

where $\widehat{\boldsymbol{\beta}}_{\mathrm{GROUM}} = \left(\boldsymbol{X}'\boldsymbol{W}\boldsymbol{\Gamma}\boldsymbol{X}\right)^{-1}\boldsymbol{X}'\boldsymbol{W}\boldsymbol{\Gamma}\boldsymbol{y}$, $\boldsymbol{\Gamma}$ is a diagonal matrix with elements ζ_h^{-1} common for all units belonging to the same stratum h. In practice, it is preferable to work with a few strata or super-strata to reduce the number of new *daf* parameters and the need for known N_h to effect stratum-specific centering in the working covariance matrix (8) of GROUM. The super-strata can be interpreted as groups of strata with similar designs. It follows that the GROUM estimator has a calibration form and the calibrated weights analogous to $w_{k,\mathrm{GREG}}$, are given by

$$w_{hk,\text{GROUM}} = w_{hk}\left(1 + \zeta_h^{-1}x'_{hk}\widehat{\lambda}_{\text{GROUM}}\right), \widehat{\lambda}_{\text{GROUM}} = (X'\Gamma WX)^{-1}(T_x - t_{x,\text{HT}}).$$

(10)

It is seen from (10) that the modification of GREG by *dafs* does not affect the asymptotic unbiasedness of GREG because $\widehat{\lambda}_{\text{GROUM}}$ continues to be $o_p(n^{-1/2})$ under usual regularity conditions as in GREG. It is easily seen that for STSRS, the optimal choice of ζ_h (assuming $n_h \cong n_h - 1$) for minimizing the true variance is given by

$$\zeta_{h,\text{STSRS}} = f_h(1 - f_h)^{-1}/\sum_{h=1}^{H} f_h(1 - f_h)^{-1},$$

(11)

where f_h is the sampling fraction n_h/N_h for stratum h.

For general designs, the optimal choice of *dafs* ζ_h's can be specified by a grid search such that the stratified version of the UWE—(a convenient substitute for the design-based generalized variance) is minimized. Specifically, UWE in this case is defined as

$$\text{UWE}_{\text{GROUM}} = \left(\sum_{h=1}^{H} n_h^{-1}(N_h/N)^2\right)^{-1}$$

$$\times \sum_{h=1}^{H}\left((N_h/N)^2\sum_{k=1}^{n_h} w_{hk,\text{GROUM}}^2/\left(\sum_{k=1}^{n_h} w_{hk,\text{GROUM}}\right)^2\right).$$

(12)

Thus GROUM captures to some extent different stratum designs through relative *dafs* and builds some optimality in the resulting estimator by suitably choosing *dafs* without introducing the instability problem of OR. This explains the terminology of optimal unbiased modification in the definition of GROUM. Moreover, it gives rise to general calibration estimators (such as generalized raking for range-restricted weight adjustment) which yield the linear adjustment factor of (10) as a special case. The underlying idea of the GROUM method was motivated from a similar use of *dafs* to account for different designs in multiple frames by Singh and Wu (2003); see next section.

3   Zero Functions and Regression Estimation for Two Frames

For simplicity we start with two frames. Generalizations to three or more frames are considered in Sect. 5. With two partially overlapping frames A and B, using standard notations, the population total parameter of interest is defined as

$$T_y = T_{ya(A)} + T_{yb(B)} + T_{yab(AB)},$$

(13)

where the disjoint domain totals $T_{ya(A)}$, $T_{yb(B)}$, and $T_{yab(AB)}$ correspond respectively to domains containing units from A only, B only, and the overlap of A and B. We have two independent samples s_A and s_B of sizes n_A and n_B respectively, one from each frame. Besides the traditional zero functions from auxiliary information from each frame, $t_{x(A),\mathrm{HT}} - T_{x(A)}$ and $t_{x(B),\mathrm{HT}} - T_{x(B)}$ with known control totals $T_{x(A)}$ and $T_{x(B)}$ which include frame population counts N_A, N_B assumed to be available, the challenge in MF arises with the availability of new nontraditional zero functions from the overlap domain because of two estimators for each selected outcome variable z as a covariate for the study variable y. The z-variables of course include the y-variables of interest as well as the domain indicator variables. The new zero functions are given by

$$ t_{zab(A),\mathrm{HT}} - t_{zab(B),\mathrm{HT}}, \ t_{zab(A),\mathrm{HT}} = \sum_{k=1}^{n_A} z_{kA} w_{kA}, \ t_{zab(B),\mathrm{HT}} = \sum_{k=1}^{n_B} z_{kB} w_{kB}, \quad (14) $$

where the two estimators of the overlap domain are obtained by using each sample separately and the HT-weights w_{kA} and w_{kB} are defined as usual. Clearly, there could be a number of new zero functions and this grows as the number of overlap domains with multiple frames grows. Two special zero functions corresponding to domain counts and domain y-totals are $\widehat{N}_{ab(A),\mathrm{HT}} - \widehat{N}_{ab(B),\mathrm{HT}}$, and $t_{yab(A)} - t_{yab(B),\mathrm{HT}}$. Most of the estimators proposed in the literature deal with only these two types of zero functions. The proposed regression formulation, however, allows for inclusion of other selected z-variables as well for increased efficiency. The extended set of predictor zero functions containing both traditional and nontraditional zero functions will be denoted by the vector $t_{x+,\mathrm{HT}} - T_{x+}$, where the notation x^+ signifies that the auxiliary variables could be the usual x's from A or B or the union of the two frames or the z-variables which are $+z$ for one frame and $-z$ for the other in the new zero functions defined by a pair of frames. Note that the control totals T_z for the new nontraditional zero functions are simply 0s.

3.1   GROUM for Multiple Frames with No Random Domain-Level Count Controls

The basic idea in combining information from multiple frames rests on the fact that the best linear combination of two unbiased HT estimators of the same population parameter can also be alternatively obtained as the optimal regression of the SMHT (defined as the simple average of the two) on the predictor zero function defined as the difference of the two estimators. This observation was exploited by Singh (1996) and Singh and Wu (1996) to apply the MR methodology to MF in which GREG-type calibration estimators are constructed by regressing the initial SMHT estimator on the extended class of predictor zero functions $t_{x+,\mathrm{HT}} - T_{x+}$ under a working covariance matrix. Later, Singh and Wu (2003) improved the choice of the working covariance matrix by using the concept of relative design adjustment factor (*daf*) instead of the relative effective sample size and a more general *maf* than the simple multiplicity adjustment used earlier in Singh and Wu (1996) who had termed them as scaling factor and combining factor respectively. Improved precision in the new estimates is realized by using a grid search to choose the optimal pair of *daf* and *maf* based on the two samples such that the generalized variance (or the UWE for simplicity) is minimized. More specifically, the initial SMHT estimator in the case of two frames considered here is given by

$$t_{y,\text{SMHT}} = t_{ya(A),\text{HT}} + t_{yb(B),\text{HT}} + \left(t_{yab(A),\text{HT}} + t_{yab(B),\text{HT}}\right)/2. \tag{15}$$

Let M denote the $(n_A + n_B)$-dimensional diagonal matrix of simple *mafs* denoted by $\alpha_{k(A)}^{\text{SM}}$ and $\alpha_{k(B)}^{\text{SM}}$ which are 1 for units in subsamples $s_{a(A)}$ and $s_{a(B)}$ and 1/2 in subsamples $s_{ab(A)}$ and $s_{ab(B)}$, and let $\boldsymbol{\Delta}$ denote the $(n_A + n_B)$-dimensional diagonal matrix of inverse of relative *dafs*, $0 < \eta_A, \eta_B < 1$, $\eta_A + \eta_B = 1$, with η_A for units in s_A and η_B for units in s_B. The relative *dafs* (η's) make adjustments in the working covariance structure for possibly different designs for samples from different frames. Here for simplicity we will not consider additional relative *dafs* (ζ_h) to capture effects of possibly different (super-) stratum-specific designs within a frame. Now the MF estimator termed GROUM(*s*) (*s* for simple multiplicity adjustment) can be defined analogous to GROUM of (9) for SF as a regression estimator given below:

$$t_{y,\text{GROUM}(s)} = t_{y,\text{SMHT}} - \widehat{\boldsymbol{\beta}}'_{\text{GROUM}(s)}\left(t_{x+,\text{HT}} - T_{x+}\right), \tag{16}$$

where $\widehat{\boldsymbol{\beta}}_{\text{GROUM}(s)} = \left(X'\boldsymbol{\Delta} W X\right)^{-1} X'\boldsymbol{\Delta} W M y$. Note that since for each unit k, the variable y can be factored out of $t_{y,\text{SMHT}}$ and $\widehat{\boldsymbol{\beta}}'_{\text{GROUM}(s)}$, the above estimator can also be obtained as a calibration estimator $\sum_{s_A} w_{k(A),\text{GROUM}(s)} y_{k(A)} + \sum_{s_B} w_{k(B),\text{GROUM}(s)} y_{k(B)}$ where

$$w_{k(A),\text{GROUM}(s)} = w_{k(A)}\left(1 + \eta_A^{-1} x_{k(A)}^{+\prime} \widehat{\boldsymbol{\lambda}}_{x+}\right),$$
$$w_{k(B),\text{GROUM}(s)} = w_{k(B)}\left(1 + \eta_B^{-1} x_{k(B)}^{+\prime} \widehat{\boldsymbol{\lambda}}_{x+}\right), \tag{17}$$

where $\widehat{\boldsymbol{\lambda}}_{x+} = \left(X'W\boldsymbol{\Delta}X\right)^{-1}\left(T_{x+} - t_{x+,\text{HT}}\right)$, the components of the parameter estimate $\widehat{\boldsymbol{\lambda}}_{x+}$ may be frame-specific depending on the corresponding covariate in the set x^+. For example, for traditional zero functions with corresponding components of T_{x+} being N_A or N_B, the respective components of $\widehat{\boldsymbol{\lambda}}_{x+}$ are either A- or B-specific. However, for the new zero functions $t_{zab(A),\text{HT}} - t_{zab(B),\text{HT}}$ with corresponding components of T_{x+} being 0, the respective components of $\widehat{\boldsymbol{\lambda}}_{x+}$ are not frame-specific.

3.2 GROUM for Multiple Frames with Random Domain-Level Count Controls

In practice, it might be better to use the count zero function $\widehat{N}_{ab(A),\text{HT}} - \widehat{N}_{ab(B),\text{HT}}$ for Hájek-ratio adjustment of the initial SMHT estimator before performing regression on the remaining zero functions because the initial estimator may not be well correlated with the count zero predictor; see the simulation results of Singh and Mecatti (2013). This can be implemented in two steps. In the first step, a best linear combination $\tilde{N}_{ab(AB)}$ of $\widehat{N}_{ab(A),\text{HT}}$ and $\widehat{N}_{ab(B),\text{HT}}$ is obtained which gives rise to three domain level random count controls: $\tilde{N}_{a(A)}$ obtained as $N_A - \tilde{N}_{ab(AB)}$, $\tilde{N}_{a(B)}$ as $N_B - \tilde{N}_{ab(AB)}$, and of course $\tilde{N}_{ab(AB)}$. The initial estimators now becomes SM$\tilde{H}$HT where $\tilde{H}$ signifies Hájek-ratio adjustment to random controls. In the second step, $t_{y,\text{SM}\tilde{H}\text{HT}}$ is regressed on t_{x+}, $\tilde{H}\text{HT} - T_{x+}$ without including N_A and N_B as components of T_{x+} as these controls continue to be satisfied after performing regression estimation to be termed GROUM(s^*) where s^* denotes that the initial estimator is SM$\tilde{H}$HT and not SMHT. An alternative simpler way to implement GROUM(s^*) is to further extend the vector of zero functions to obtain $t_{x^*,\text{HT}} - T_{x^*}$ which includes new random domain-level controls $\tilde{N}_{a(A)}$, $\tilde{N}_{a(B)}$, and $\tilde{N}_{ab(AB)}$ in place of the original fixed controls N_A and N_B (in fact, original fixed controls are automatically implied by the new random domain control counts) and where indicator vectors for the three domain samples are now used as x-variables in place of the indicators for units in the full samples s_A and s_B. It turns out that the initial transformation of SMHT to SM$\tilde{H}$HT is no longer needed. It is easily shown that the GROUM(s^*) estimator is given by

$$t_{y,\text{GROUM}(s*)} = t_{y,\text{SMHT}} - \widehat{\beta}'_{\text{GROUM}(s*)} \left(t_{x*,\text{HT}} - T_{x*} \right). \tag{18}$$

The η-parameters are specified by minimizing the UWE of GROUM(s^*) calibrated weights via a grid search. That is, for each $\eta_A = 0.1, 0.2, \ldots, 0.9$, GROUM($s^*$) calibrated weights are computed along with the corresponding UWE. The optimal choice of η_A corresponds to the smallest UWE observed. The UWE for GROUM(s^*) analogous to (12) can be computed by treating the three domains as strata and is defined as

$$\text{UWE}_{\text{GROUM}(s*)} = \left(F_{a(A)}^{(c)} + F_{b(B)}^{(c)} + F_{ab(A)}^{(c)} + F_{ab(B)}^{(c)} \right) \times$$
$$\left(F_{a(A)}^{(s)} + F_{b(B)}^{(s)} + F_{ab(A)}^{(s)} + F_{ab(B)}^{(s)} \right)^{-1}, \tag{19}$$

where the superscript "c" for various terms signify a complex design and "s" signifies a simple design (i.e., SRS), and terms in the numerator are $F_{a(A)}^{(c)} = \left(\tilde{N}_{a(A)} / \widehat{N}_{a(A)}^* \right)^2 \sum_{s_{a(A)}} w_{k(A)}^{*2}$, $\widehat{N}_{a(A)}^* = \sum_{s_{a(A)}} w_{k(A)}^*$, $w_{k(A)}^*$ as an abbreviation for $w_{k(A),\text{GROUM}(s*)}$; $F_{ab(A)}^{(c)} = \left(\alpha_A^{\text{SM}}\right)^2 \left(\tilde{N}_{ab(AB)} \big/ \widehat{N}_{ab(A)}^*\right)^2 \sum_{s_{ab(A)}} w_{k(A)}^{*2}$, and all other terms are defined in an analogous manner. Terms in the denominator are $F_{a(A)}^{(s)} = \tilde{N}_{a(A)}^2 \big/ n_{a(A)}$, $F_{ab(A)}^{(s)} = \left(\alpha_A^{\text{SM}}\right)^2 \tilde{N}_{ab(A)}^2 \big/ n_{ab(A)}$, other terms being analogous. Note that the expression (19) simplifies considerably for GROUM($s*$) because $\widehat{N}_{a(A)}^* = \tilde{N}_{a(A)}$ along with similar equalities for other terms due to calibration, and $\alpha_A^{\text{SM}} = \alpha_{kA}^{\text{SM}} = 1/2$ for all k in $s_{ab(A)}$.

It may be observed that the above estimator is not invariant to the choice of *mafs*. If we choose general *mafs* ($\alpha_{k(A)}^{\text{GM}}$ and $\alpha_{k(B)}^{\text{GM}}$) for the initial estimator (denoted by $t_{y,\text{GMHT}}$) before regression as in (16), then we get a new estimator to be termed GROUM(g). We can set $\alpha_{k(A)}^{\text{GM}}$ at a common value of α_A^{GM} for all k in $s_{ab(A)}$ and similarly α_B^{GM} for $s_{ab(B)}$ where we need $\alpha_A^{\text{GM}} + \alpha_B^{\text{GM}} = 1$ for unbiasedness, then we can find the optimal pair ($\alpha_A^{\text{GM}}, \eta_A$) by minimizing the new UWE as before. The estimator GROUM($g*$) can also be defined analogous to GROUM($s*$). With three or more frames, however, the number of disjoint overlap domains may be large, and the task of finding an optimal *maf* for each overlap domain may be onerous along with an optimal *daf* for each frame. In such circumstances, it may be adequate to work with the SMHT estimator as the initial one. As mentioned earlier, the above GROUM formulation for MF easily allows for incorporating many more zero functions $t_{zab(A),\text{HT}} - t_{zab(B),\text{HT}}$ corresponding to different z-variables although typically the zero function corresponding to the study variable y is considered in the literature.

4   Calibration Adjustment Factor as a Coverage Bias Model

In regression estimation for SF or MF considered so far, coverage bias was assumed to be absent and the goal of regression on predictor zero functions was to reduce variance. However, in practice, coverage bias is invariably present and fortunately, calibration methods used for variance reduction also work in favor of bias reduction. Thus calibration methods have a dual property; see Folsom and Singh (2000), Singh and Folsom (2000), Kott (2006), and Särndal (2007). The reason for this happy coincidence is that the calibration adjustment factor can also be justified as a coverage bias adjustment factor under a super population model similar to nonresponse bias models. We discuss it further for SF and MF in the following.

4.1   Coverage Bias Models for Single Frames

The GREG estimator (6) can also be obtained as a calibration estimator under a super-population coverage bias model for the weight adjustment factor $a_{k,\text{GREG}}(\boldsymbol{\lambda})$; compare it with (7). We define for the kth unit in the finite population,

$$a_{k,\text{GREG}}(\lambda) = 1 + x_k'\lambda, \tag{20}$$

where the covariates used for variance reduction via regression are also deemed good predictors for coverage. The elements of the p-dimensional vector parameter λ can take arbitrary real values. As explained in Folsom and Singh (2000), the inverse of $a_{k,\text{GREG}}(\lambda)$ is assumed to model the expected number of times a unit is listed in the imperfect frame for the population under consideration. This expected value should be nonnegative by definition and a value less than 1 implies undercoverage while a value more than 1 implies overcoverage with the reverse implication for the adjustment factor. The linear model (20) in itself does not guarantee positive adjustment factors but it generally works well in practice. A more suitable model will be to use a log linear model for the expected number of time a unit is listed and then the adjustment factor is given by $a_k(\lambda) = \exp(x_k'\lambda)$ which is also obtained under the raking-ratio method of calibration estimation or the calibration class of generalized raking estimators of Deville and Särndal (1992). Note that for small λ, $\exp(x_k'\lambda) \cong 1 + x_k'\lambda$, the linear adjustment factor under GREG.

Similar to GREG, the GROUM estimator (9) can also be obtained as a calibration estimator under the following super-population coverage bias model for the weight adjustment factor $a_{k,\text{GROUM}}(\lambda)$; compare it with (10). We have for the kth unit in stratum h subpopulation,

$$a_{hk,\text{GROUM}}(\lambda) = 1 + \zeta_h^{-1}x_{hk}'\lambda, \quad \sum_{h=1}^{H}\zeta_h = 1, \quad 0 < \zeta_h < 1, \tag{21}$$

where the factor ζ_h can no longer be interpreted as *daf* because it is meaningless at the super-population level in the absence of any sampling design at that stage. However, it can be interpreted as a factor that captures differential coverage bias phenomena across different strata. If it is set close to 0, it inflates relatively the impact of the parameters λ for bias adjustment in any given stratum but if it is set close to 1, it deflates the impact of bias in that stratum relative to others. Like the linear model (20), in order to be meaningful as inverse of the expected number of times a unit is listed, the adjustment factor of (21) should be positive. This restriction can be easily achieved by considering a nonlinear model given by $a_{hk}(\lambda) = \exp(\zeta_h^{-1}x_{hk}'\lambda)$ which is indeed close to the linear model (21) for small λ.

The bias adjustment model (20) or (21) is only specified under first moment assumptions and can be fit using the method of moments resulting in calibration estimating equations. The basic moment condition is that the HT estimators of x-variables after bias adjustments must equal the true control totals T_x; i.e., the p zero functions $t_{x,\text{HT}} - T_x$ become actually zero after bias adjustments. Thus the estimating equations for the parameters λ using the calibration method for model (21) are given by

$$\sum_{h=1}^{H}\sum_{k=1}^{n_h} x_{hk}w_{hk}\left(1 + \zeta_h^{-1}x_{hk}'\lambda\right) = T_x \tag{22}$$

whose solution is identical to $\widehat{\lambda}_{\text{GROUM}} = \left(X'\Gamma WX\right)^{-1}\left(T_x - t_{x,\text{HT}}\right)$ of (10). Therefore, the resulting calibration estimator for coverage bias reduction is the same as the GROUM estimator (9) constructed for variance reduction. Similar results are obtained for the GREG estimator.

4.2   Coverage Bias Models for Multiple Frames

The modeling here is analogous to the bias adjustment model (21) for GROUM. For the two frames we assume that the superpopulation coverage bias models for the weight adjustment factor are defined as

$$a_{k(A)}\left(\lambda_{x+}\right) = \left(1 + \eta_A^{-1}x_{k(A)}^{+/}\lambda_{x+}\right), \quad a_{k(B)}\left(\lambda_{x+}\right) = \left(1 + \eta_B^{-1}x_{k(B)}^{+/}\lambda_{x+}\right),$$

$$0 < \eta_A, \quad \eta_B < 1, \quad \eta_A + \eta_B = 1, \tag{23}$$

where covariates x^+ chosen correspond to either the GROUM(s) or GROUM(g) estimator. As before, we can use log linear models for the inverse adjustment factor to ensure desirable positive values of the adjustment factor. The relative *dafs* η_A, η_B should be interpreted free from design considerations at the super-population level. They account for differential frame-specific coverage bias, and relatively small values inflate the impact of the model parameters in a given frame relative to other frames while relatively large values deflate the impact of model parameters in that frame. Analogous to (22) the estimating equations for λ_{x+} using the calibration method for MF are given by

$$\alpha_A^{\text{SM}}\sum_{s_A} x_{k(A)}^+ w_{k(A)}\left(1 + \eta_A^{-1}x_{k(A)}^{+/}\lambda_{x+}\right)$$

$$+\alpha_B^{\text{SM}}\sum_{s_B} x_{k(B)}^+ w_{k(B)}\left(1 + \eta_B^{-1}x_{k(B)}^{+/}\lambda_{x+}\right) = T_{x+}, \tag{24}$$

where *mafs* (α's) are required to sum to 1 for unbiasedness of above estimating equations. In the above equation we used simple *mafs* which yield the estimate of λ_{x+} identical to $\widehat{\lambda}_{x+} = \left(X'W\Delta X\right)^{-1}\left(T_{x+} - t_{x+,\text{HT}}\right)$ of (17), and hence the resulting calibration estimator after bias adjustments is identical to the GROUM(s) estimator as desired. Similarly other estimators GROUM(s^*), GROUM(g), and GROUM(g^*) can be obtained as bias adjusted calibration estimators.

5 Conclusions

In this chapter we reviewed use of zero functions and GREG-type regression calibration methods for estimating parameters from SF and MF. This included GROUM developed as an enhancement of GREG for building optimality and multiplicity adjustments for MF. For MF, the general unified framework based on generalized multiplicity-adjusted Horvitz–Thompson (GMHT) estimators and regression on predictor zero functions gives rise to new and improved estimators in the class of calibration estimators; see Singh and Mecatti (2011, 2013) for more details. Although in this chapter we focused on GMHT-regression calibration estimators, various existing MF methods available in the literature (using either the separate or the combined frame approach) can be shown to belong to the GMHT-Regression class (with or without a calibration form in general) as shown in Singh and Mecatti (2013). It is remarked that regression calibration of GMHT estimators in MF is analogous to regression calibration of HT estimators in SF. Similarly variance estimation of GMHT-regression estimators is analogous to HT-regression because they can be expressed as sums of contributions from different frames. To see this for the general MF estimation problem with Q frames, we consider, for example, the GROUM(g) estimator for Q frames as follows:

$$t_{y,\text{GROUM}(g)} = t_{y,\text{SMHT}} - \widehat{\boldsymbol{\beta}}'_{\text{GROUM}(g)} \left(\boldsymbol{t}_{x+,\text{HT}} - \boldsymbol{T}_{x+} \right), \tag{25a}$$

$$= \sum_{q=1}^{Q} \sum_{k=1}^{n_q} \alpha_q^{\text{GM}} w_{k(q)} y_{k(q)} - \sum_{i=1}^{p+} \widehat{\beta}_i \left(\sum_{q'=1}^{Q} \delta_{i(q')} \left(t_{x_i^+(q'),\text{HT}} - T_{x_i^+} \right) \right), \tag{25b}$$

where $\widehat{\beta}_i$ is the ith element of $\widehat{\boldsymbol{\beta}}'_{\text{GROUM}(g)}$, $i = 1$ to p^+, p^+ being the number of x-variables in x^+. The ith element of $\boldsymbol{t}_{x+,\text{HT}} - \boldsymbol{T}_{x+}$ is a predictor zero function which is either based on a single frame if it is of the traditional type of zero function or two frames if it is nontraditional in which case $T_{x_i^+}$ assumes the value of zero. In general, we can express it as a sum $\sum_{q'=1}^{Q} \delta_{i(q')} \left(t_{x_i^+(q'),\text{HT}} - T_{x_i^+} \right)$ where $\delta_{i(q')}$ is 0 if frame q' does not contribute to the ith element, -1 if frame q' contributes but appears as the second term with a negative sign in the nontraditional zero function formed by a pair of overlapping frames, and $+1$ if it contributes but appears as the first term with a positive sign in the nontraditional zero function with a zero control total or in the traditional zero function with a nonzero control total. It is easily seen from the above general form (25b) of the MF estimator that if $\widehat{\beta}_i$ is computed under a working covariance structure such that the y value can be factored out, then we can write it as a calibration estimator, as is indeed the case with GROUM(g). For variance estimation using standard HT formulae, $t_{y,\text{GROUM}(g)}$ for example, can be expressed as

$$t_{y,\text{GROUM}(g)} = \sum_q \left(\sum_{s_q} \left(\alpha_q^{\text{GM}} y_{k(q)} - \sum_{i=1}^{p+} \widehat{\beta}_i \delta_{i(q)} x_{ik(q)}^+ \right) w_{k(q)} \right) + \text{const.,} \quad (26)$$

from which a variance estimate of $t_{y,\text{GROUM}(g)}$ can be easily obtained using independence of samples from different frames. In the case of GROUM(g^*), however, since random domain-level count controls are used, the variance needs to be adjusted for extra variability due to these random controls. This implies that after linearization of random controls, certain extra frame-specific terms corresponding to frames involved in computing the random control need to be included in the regression residual form for GROUM(g^*) analogous to (26).

Finally we remark that in practice for estimation with MF data, invariably nonresponse bias adjustments also need to be made before any coverage bias adjustments. For SF, use of calibration estimation with generalized exponential models (GEM, also known as the WTADJUST procedure in SUDAAN V10)) for modeling nonresponse bias and coverage bias was proposed by Folsom and Singh (2000) such that the weight adjustment factor can be range-restricted (e.g., greater than 1 for nonresponse bias and greater than 0 for coverage bias) analogous to the generalized raking calibration methods of Deville and Särndal (1992) proposed for variance reduction. The GROUM method discussed here for coverage bias adjustment is an improvement over GEM (in view of the built-in variance optimality through *dafs* and model-based smoothing of extreme weights instead of the traditional ad-hoc method of weight trimming although the latter was not discussed here) can also be easily used for nonresponse bias adjustments with MF via a calibration approach with suitable range restrictions on weight adjustments.

References

Binder, D.A.: On the variances of asymptotically normal distributed estimators from complex surveys. Int. Stat. Rev. **51**, 279–292 (1983)

Deville, J.-C., Särndal, C.-E.: Calibration estimators in survey sampling. J. Am. Stat. Assoc. **87**, 376–382 (1992)

Folsom, R.E. Jr., Singh, A.C.: A generalized exponential model for sampling weight calibration for a unified approach to nonresponse, poststratification, and extreme weight adjustments. In: Proceedings of the American Statistical Association, Survey Research Methods Section, pp. 598–603 (2000)

Fuller, W.A.: Regression analysis for sample surveys. Sankhya Ser. C **37**, 117–132 (1975)

Fuller, W.A., Rao, J.N.K.: A regression composite estimator with application to the Canadian labour force. Surv. Methodol. **27**, 45–51 (2001)

Godambe, V.P., Thompson, M.E.: Parameters of super-population and survey population: their relationships and estimation. Int. Stat. Rev. **54**(2), 127–138 (1986)

Godambe, V.P., Thompson, M.E.: An extension of quasi-likelihood estimation (with discussion). J. Stat. Plan. Res. **12**, 137–172 (1989)

Godambe, V.P., Thompson, M.E.: Estimating functions and survey sampling. In: Pfeffermann, D., Rao, C.R. (eds.) Handbook of Statistics: Sample Surveys: Inference and Analysis, vol. 29B, pp. 83–101. Elsevier, North-Holland (2009)

Kott, P.A.: Using calibration weighting to adjust for nonresponse and coverage errors. Surv. Methodol. **33**, 133–144 (2006)

McCullagh, P., Nelder, J.A.: Generalized Linear Models, 2nd edn. Chapman and Hall, London (1989)

Rao, J.N.K.: Estimating totals and distribution functions using auxiliary information at the estimation stage. J. Official Stat. **10**, 153–165 (1994)

Särndal, C.-E.: On π-inverse weighting versus best linear unbiased weighting in probability sampling. Biometrika **67**, 639–650 (1980)

Särndal, C.-E.: The calibration approach in survey theory and practice. Surv. Methodol. **33**, 99–119 (2007)

Singh, A.C.: Categorical data analysis for simple and complex surveys. In: Pfeffermann, D., Rao, C.R. (eds.) Handbook of Statistics: Sample Surveys: Inference and Analysis, vol. 29B, pp. 329–370. Elsevier, North-Holland (2009)

Singh, A.C.: Combining information in survey sampling by modified regression. In: Proceedings of the Survey Research Methods Section, American Statistical Association, pp. 120–129 (1996)

Singh, A.C., Folsom, R.E. Jr.: Bias corrected estimating functions approach for variance estimation adjusted for poststratification. In: Proceedings of the American Statistical Association, Section on Survey Research Methods, pp. 610–615 (2000)

Singh, A.C., Mecatti, F.: Generalized multiplicity-adjusted Horvitz–Thompson-type estimation as a unified approach to multiple frame surveys. J. Official Stat. **27**(4), 633–650 (2011)

Singh, A.C., Mecatti, F.: Generalized multiplicity-adjusted multiple frame estimation via regression. In: Proceedings of the Statistics Canada Symposium on Producing Reliable Estimates from Imperfect Frames, Ottawa (2013)

Singh, A.C., Rao, R.P.: Optimal instrumental variable estimation for linear models with stochastic regressors using estimating functions. In: Basawa, I.V., Godambe, V.P., Taylor, R.L. (eds.) Selected Proceedings of the Symposium on Estimating Functions. IMS Lecture Notes-Monograph Series, vol. 32, pp. 177–192. Institute of Mathematical Statistics, California, Hayward (1997)

Singh, A.C., Wu, S.: Estimation for multiframe complex surveys by modified regression. In: Proceedings of the Statistical Society of Canada, Section on Survey Methods, pp. 69–77 (1996)

Singh, A.C., Wu, S.: An extension of generalized regression estimator to dual frame survey. In: Proceedings of the American Statistical Association, Section on Survey Research Methods, pp. 3911–3918 (2003)

Singh, A.C., Kennedy, B., Wu, S.: Composite estimation for the Canadian labour force survey with a rotating panel design. Surv. Methodol. **27**, 33–44 (2001)

Singh, A.C., Ganesh, N., Lin, Y.: Improved sampling weight calibration by generalized raking with optimal unbiased modification. In: Proceedings of the American Statistical Association, Survey Research Methods Section, pp. 3572–3583 (2013)

Skinner, C.J.: Domain means, regression, and multivariate analysis. In: Skinner, C.J., Holt, D., Smith, T.M.F. (eds.) Analysis of Complex Surveys, pp. 59–88. Wiley, New York (1989)

The Analysis of Survey Data Using the Bootstrap

Jean-François Beaumont

Abstract We first review bootstrap variance estimation for estimators of finite population quantities such as population totals or means. In this context, the bootstrap is typically implemented by producing a set of bootstrap design weights that account for the variability due to sample selection. Sometimes, survey analysts are interested in making inferences about model parameters. We then describe how to modify bootstrap design weights so as to account for the variability resulting from the analyst's model. Finally, we discuss bootstrap tests of hypotheses for survey data.

Keywords Bootstrap test of hypotheses • Bootstrap variance estimation • Bootstrap weights • Estimating equations

1 Introduction

Survey analysts, or other users of survey data, are often interested in making inferences about finite population quantities, such as population totals or means, using sample data. The sample is selected from the finite population according to a probabilistic sampling design. Survey estimates are typically computed by weighting each sample unit by its design weight. The design weight of a given sample unit is simply the inverse of its selection probability.

In the design-based framework to inference, the variability of an estimator is due to sample selection. The variance of an estimator, often called the design variance, can be estimated using a number of methods. Linearization methods are often used in surveys where the population quantities to be estimated are known in

J.-F. Beaumont (✉)
Statistical Research and Innovation Division, Statistics Canada, Ottawa, ON, Canada
e-mail: Jean-Francois.Beaumont@statcan.gc.ca

F. Mecatti et al. (eds.), *Contributions to Sampling Statistics*, Contributions to Statistics, 53
DOI 10.1007/978-3-319-05320-2_4,
© Springer International Publishing Switzerland 2014

advance and are not too complex. The main drawback of linearization methods is
that each population quantity (e.g., a total, a mean, a ratio, a median, etc.) leads
to a different variance expression and variance estimator. Also, calibration and
nonresponse weighting adjustments may yield very complex variance expressions
that are not always easy to implement for ultimate users if appropriate software is
not available. For these reasons, replication methods may be preferred, especially
when a public-use microdata file is produced by the statistical agency and made
available to users. Replication methods, such as the jackknife or the bootstrap,
are typically implemented in practice by producing a large number of replication
weights for each sample unit. Users can then simply compute many replicates of
their original estimate, with each replication estimate obtained using a different set
of weights. The design variance estimate is then computed from the variability of
the replication estimates. This approach is convenient for users as it gives them
a simple way of computing variance estimates for different types of population
quantities. The jackknife was used at Statistics Canada for many years but it has
slowly been replaced by the bootstrap over the last 15 years, mainly because the
jackknife requires as many replication weights as there are primary sampling units.
This number may be very large in some surveys, which may raise computational
and storage issues. The bootstrap has the advantage of allowing the user to control
the number of replication weights. It is also known to be more accurate than the
jackknife for non-smooth estimators such as the estimator of a population median.
In the remaining of this chapter, we will focus on the bootstrap and on the use of
bootstrap design weights (e.g., Beaumont and Patak 2012; Rao et al. 1992) as a
means of taking care of the sampling variability.

Analysts may not only be interested in making inferences about finite population
quantities, they may also be interested in studying relationships among variables. To
this end, they may postulate and validate a model and try to draw conclusions about
model parameters based on sample data. When analysts are interested in making
inferences about model parameters, two sources of variability are normally taken
into account: the model that generates data of the finite population and the sampling
design. When the overall sampling fraction is negligible, the model variability can
be ignored and standard bootstrap techniques that account for the sampling design
variability can be used. However, there are many practical cases where the model
variability cannot be ignored. We show how to modify bootstrap design weights in
a simple way so as to account for the model variability (see Beaumont and Charest
2012). The survey analyst may also be interested in testing hypotheses about model
parameters. This can be achieved by replicating a simple weighted model-based test
statistic using the bootstrap weights (see Beaumont and Bocci 2009).

In Sect. 2, we review the bootstrap for design variance estimation when finite
population quantities are of interest. In Sect. 3, we show how to modify bootstrap
design weights to account for the model variability in the context of inference for
model parameters. We also discuss bootstrap tests of hypotheses. We provide a short
conclusion in the last section.

2 Bootstrap for Finite Population Quantities

Let us consider the estimation of the total $T_y = \sum_{k \in U} y_k$ of a finite population U, where y_k is the value of the variable of interest y for unit k. A sample s is selected from the finite population U using a sampling design $p(s)$ and the variable of interest is collected only for sample units. The expansion estimator of the total T_y is $\widehat{T}_y = \sum_{k \in s} w_k y_k$, where $w_k = \pi_k^{-1}$ is the design weight of unit k and $\pi_k = \Pr(k \in s)$ is the probability that unit k is selected in the sample. The expansion estimator is p-unbiased (design-unbiased) for T_y, i.e., $E_p\left(\widehat{T}_y\right) = T_y$, and it is assumed to be p-consistent (design-consistent) in the sense that $\widehat{T}_y - T_y = O_p\left(N/\sqrt{n}\right)$; e.g., see Fuller (2009, Chapter 1). The subscript p refers to the sampling design. Design consistency is achieved when the design variance of the expansion estimator, $\text{var}_p\left(\widehat{T}_y\right)$, is $O(N^2/n)$. Informally speaking, this means that, as the sample and population size increase, $\text{var}_p\left(\widehat{T}_y\right)$ does not become too far from the variance that would be obtained under simple random sampling.

We also assume that the design variance, $\text{var}_p\left(\widehat{T}_y\right)$, can be estimated through an estimator that takes the following quadratic form:

$$v(y) = \sum_{k \in s} \sum_{l \in s} \sigma_{kl} y_k y_l. \tag{1}$$

The quantity σ_{kl} is usually chosen so that $v(y)$ is a p-unbiased estimator of $\text{var}_p\left(\widehat{T}_y\right)$. We assume that $v(y)$ is also p-consistent for $\text{var}_p\left(\widehat{T}_y\right)$ in the sense that $v(y)/\text{var}_p\left(\widehat{T}_y\right)$ converges to 1 in design probability. For instance, the Horvitz–Thompson estimator of $\text{var}_p\left(\widehat{T}_y\right)$ is obtained from (1) using $\sigma_{kl} = w_k w_l - \pi_{kl}^{-1}$, where $\pi_{kl} = \Pr(k \in s, l \in s)$ is the probability that both units k and l are selected in the sample. For simplicity, we will assume that $v(y)$ cannot take negative values. Beaumont and Patak (2012) point out that the Sen-Yates-Grundy variance estimator, which can be written in the form (1), is non-negative for fixed-size sampling designs provided that $\pi_{kl} - \pi_k \pi_l \leq 0$ for all $k \neq l$. They also discuss in their conclusion how to fix a variance estimator that can take negative values.

Most bootstrap methods in the literature can be implemented by computing bootstrap design weights, denoted by w_k^*, $k \in s$. The expression for the bootstrap weights is given explicitly in Rao et al. (1992), Bertail and Combris (1997), Chipperfield and Preston (2007), Antal and Tillé (2011), and Beaumont and Patak (2012) but can also be derived with most other methods. This is the case for the bootstrap methods described in the review paper by Sitter (1992) and those based on the creation of a pseudo-population such as Gross (1980), Chao and Lo (1985), Booth et al. (1994) and Chauvet (2007). For any given sample s, the bootstrap design weights w_k^*, $k \in s$, are random variables that are generated under a certain multivariate distribution, called the bootstrap mechanism.

Let us denote by $\widehat{T}_y^* = \sum_{k \in s} w_k^* y_k$, the bootstrap replicate of $\widehat{T}_y = \sum_{k \in s} w_k y_k$. Valid bootstrap variance estimation methods must at least satisfy the following two conditions:

(1) $E_*\left(\widehat{T}_y^*\right)$ is p-consistent for $E_p\left(\widehat{T}_y\right)$; and

(2) $\text{var}_*\left(\widehat{T}_y^*\right)$ is p-consistent for $\text{var}_p\left(\widehat{T}_y\right)$.

The subscript * indicates that the expectation and variance are evaluated with respect to the bootstrap mechanism conditionally on s. Conditions (1) and (2) means that the first two bootstrap moments of $\widehat{T}_y^*$ must capture the first two design moments of $\widehat{T}_y$. These two conditions are satisfied for all the bootstrap methods cited above.

One natural way of satisfying conditions (1) and (2) is to define the bootstrap mechanism so that:

(3) $E_*\left(\widehat{T}_y^*\right) = \widehat{T}_y$; and

(4) $\text{var}_*\left(\widehat{T}_y^*\right) = v(y)$.

It is then straightforward to see that conditions (3) and (4) are satisfied if, for all $k, l \in s$:

(5) $E_*(w_k^*) = w_k$; and

(6) $\text{cov}_*(w_k^*, w_l^*) = \sigma_{kl}$.

This means that a bootstrap method is valid provided that it can be implemented through a bootstrap mechanism that satisfies conditions (5) and (6). This provides a unifying view of bootstrap methods proposed in the literature (see also Bertail and Combris 1997; Antal and Tillé 2011; Beaumont and Patak 2012). This bootstrap methodology is thus applicable to most sampling designs used in practice as long as the first-order and second-order selection probabilities can be computed, or at least, accurately approximated.

For each given sampling design, there are many possible multivariate distributions for the bootstrap design weights that satisfy the first two moment conditions (5) and (6). Beaumont and Patak (2012) provided a few general choices and propose a specific one for Poisson sampling that captures the third design moment as well (see also Bertail and Combris 1997; Antal and Tillé 2011, for other choices). For stratified simple random sampling designs, it is worth mentioning that the popular Rao and Wu (1988) method is a special case of the above general bootstrap methodology. These authors did not provide any expression for the bootstrap design weights but it is not too difficult to extract them from their equations, as shown in Beaumont and Patak (2012). Later on, Rao et al. (1992) gave an expression for the bootstrap design weights under stratified multistage sampling designs assuming with-replacement sampling at the first stage. An extension of the above bootstrap methodology to two-phase sampling was considered in Langlet et al. (2008) and applied to a few postcensal surveys in Canada. Each bootstrap method found in the literature corresponds to a certain multivariate distribution for the bootstrap design weights. The key property is to satisfy conditions (1) and (2) or conditions (5) and (6).

The expansion estimator $\widehat{T}_y$ is linear in the sense that it can be written as a linear function of the sample selection indicators I_k, $k \in U$; i.e., $\widehat{T}_y = \sum_{k \in U} I_k w_k y_k$, where $I_k = 1$, if $k \in s$, and $I_k = 0$, otherwise. In practice, most estimators of interest are nonlinear. For instance, the user might be interested in estimating the ratio $\sum_{k \in U} y_{1k} / \sum_{k \in U} y_{2k}$ by the estimator $\sum_{k \in s} w_k y_{1k} / \sum_{k \in s} w_k y_{2k}$, where y_1 and y_2 are two different variables of interest. More generally, denote by $\mathbf{y}$, a vector of variables of interest and by $\mathbf{T}_y = \sum_{k \in U} \mathbf{y}_k$, the corresponding vector of population totals. Let us also suppose that the estimator $\widehat{\Delta} = g\left(\widehat{\mathbf{T}}_\mathbf{y}\right)$ is used to estimate the finite population quantity $\Delta = g(\mathbf{T}_y)$, where $\widehat{\mathbf{T}}_\mathbf{y} = \sum_{k \in s} w_k \mathbf{y}_k$ and $g(\cdot)$ is a smooth function. It can be shown through standard Taylor linearization that, for large samples, $\widehat{\Delta}$ can be approximated in the linear form:

$$\widehat{\Delta} \approx h\left(\mathbf{T}_\mathbf{y}\right) + \sum_{k \in s} w_k z_k\left(\mathbf{T}_\mathbf{y}\right),$$

where $h(\mathbf{T}_\mathbf{y}) = g(\mathbf{T}_\mathbf{y}) - g'(\mathbf{T}_\mathbf{y})\mathbf{T}_\mathbf{y}$, $z_k(\mathbf{T}_\mathbf{y}) = g'(\mathbf{T}_\mathbf{y})\mathbf{y}_k$ and $g'\left(\widetilde{\mathbf{T}}_\mathbf{y}\right) = \frac{\partial g(\widetilde{\mathbf{T}}_\mathbf{y})}{\partial \widetilde{\mathbf{T}}_\mathbf{y}^\mathrm{T}}$, for some vector $\widetilde{\mathbf{T}}_\mathbf{y}$. In general, we thus have that $\mathrm{var}_p\left(\widehat{\Delta}\right) \approx \mathrm{var}_p\left(\sum_{k \in s} w_k z_k\left(\mathbf{T}_\mathbf{y}\right)\right)$. A standard linearization estimator of the design variance $\mathrm{var}_p\left(\widehat{\Delta}\right)$ is simply $v\left\{z\left(\widehat{\mathbf{T}}_\mathbf{y}\right)\right\}$, with the function $v(\cdot)$ defined in equation (1). The bootstrap replicate of $\widehat{\Delta}$ is $\widehat{\Delta}^* = g\left(\widehat{\mathbf{T}}_\mathbf{y}^*\right)$, where $\widehat{\mathbf{T}}_\mathbf{y}^* = \sum_{k \in s} w_k^* \mathbf{y}_k$. We can again use Taylor linearization to show that, for large samples, $\widehat{\Delta}^*$ can be approximated as

$$\widehat{\Delta}^* \approx h\left(\widehat{\mathbf{T}}_\mathbf{y}\right) + \sum_{k \in s} w_k^* z_k\left(\widehat{\mathbf{T}}_\mathbf{y}\right).$$

As a result, we normally have that $\mathrm{var}_*\left(\widehat{\Delta}^*\right) \approx \mathrm{var}_*\left(\sum_{k \in s} w_k^* z_k\left(\widehat{\mathbf{T}}_\mathbf{y}\right)\right)$. Under conditions (5) and (6), the latter reduces to $\mathrm{var}_*\left(\widehat{\Delta}^*\right) \approx v\left\{z\left(\widehat{\mathbf{T}}_\mathbf{y}\right)\right\}$. This means that, for large samples, the bootstrap variance $\mathrm{var}_*\left(\widehat{\Delta}^*\right)$ is close to the linearization variance estimator $v\left\{z\left(\widehat{\mathbf{T}}_\mathbf{y}\right)\right\}$. Therefore, if $v\left\{z\left(\widehat{\mathbf{T}}_\mathbf{y}\right)\right\}$ is thought to be a valid estimator of $\mathrm{var}_p\left(\widehat{\Delta}\right)$ then a bootstrap method that satisfies (5) and (6) should also be valid. The case of nonlinear estimators defined through estimating equations is discussed in Beaumont and Patak (2012).

Unlike the linear case, the bootstrap variance of a nonlinear bootstrap estimator, e.g. $\widehat{\Delta}^* = g\left(\widehat{\mathbf{T}}^*_\mathbf{y}\right)$, cannot be computed exactly in general. It is usually approximated through Monte Carlo simulations by generating independently many replicates of the bootstrap design weights, say B, using the same bootstrap mechanism. These bootstrap design weights, $w_k^{*(b)}, k \in s$ and $b = 1, \ldots, B$, are then used to obtain B bootstrap replicates, $\widehat{\Delta}^{*(b)}, b = 1, \ldots, B$, of the initial estimate $\widehat{\Delta}$. The bootstrap variance, $\mathrm{var}_*\left(\widehat{\Delta}^*\right)$, is approximated through the variability of these B bootstrap replicates using $B^{-1}\sum_{b=1}^{B}\left(\widehat{\Delta}^{*(b)} - \widehat{\Delta}\right)^2$.

3   Bootstrap for Inferences About Model Parameters

The survey analyst may want to assume that the data of the finite population U are generated according to a certain model: $F(y_k|\mathbf{x}_k; \boldsymbol{\beta}, \boldsymbol{\theta})$. The model describes the assumed relationship between the dependent variable, y_k, and a vector of explanatory variables, $\mathbf{x}_k, k \in U$. The vectors $\boldsymbol{\beta}$ and $\boldsymbol{\theta}$ are unknown model parameters and the analyst is interested in making inferences only about $\boldsymbol{\beta}$. We suppose further that the analyst assumes that the observations $y_k, k \in U$, are independent under the model. A typical example is the linear regression model for which $E_m(y_k|\mathbf{x}_k) = \mathbf{x}'_k\boldsymbol{\beta}$ and $\mathrm{var}_m(y_k|\mathbf{x}_k) = \theta$. The subscript m indicates that the expectation and variance are evaluated with respect to the model. From now on, we will omit conditioning on $\mathbf{x}_k$ to simplify notation when taking model expectations and variances. For instance, we will write $E_m(y_k)$ instead of $E_m(y_k|\mathbf{x}_k)$.

If a census could be conducted, the unknown model parameter $\boldsymbol{\beta}$ could be estimated by $\widehat{\boldsymbol{\beta}}_U$ through the estimating equation $\mathbf{S}\left(\widehat{\boldsymbol{\beta}}_U\right) \equiv \sum_{k \in U} \mathbf{S}_k\left(\widehat{\boldsymbol{\beta}}_U\right) = \mathbf{0}$, where $\mathbf{S}_k(\cdot)$ is an estimating function such that $\mathbf{S}(\cdot)$ is an m-unbiased estimating function for $\boldsymbol{\beta}$ in the sense that $E_m(\mathbf{S}(\boldsymbol{\beta})) = \mathbf{0}$. For the above linear regression model, the standard estimating function is: $\mathbf{S}_k(\boldsymbol{\beta}) = (y_k - \mathbf{x}'_k\boldsymbol{\beta})\mathbf{x}_k$.

As described in Sect. 2, a sample s is selected from the finite population U so that $\widehat{\boldsymbol{\beta}}_U$ cannot be computed. The most common way to handle this issue in practice is to replace $\mathbf{S}(\boldsymbol{\beta})$ by the weighted estimating function $\mathbf{S}_w(\boldsymbol{\beta}) = \sum_{k \in s} w_k \mathbf{S}_k(\boldsymbol{\beta})$. The unknown model parameter $\boldsymbol{\beta}$ can then be estimated by $\widehat{\boldsymbol{\beta}}_{ws}$ through the estimating equation $\mathbf{S}_w\left(\widehat{\boldsymbol{\beta}}_{ws}\right) = \mathbf{0}$ (e.g., Binder 1983; Rao et al. 2002; Demnati and Rao 2010). The weighted estimating function $\mathbf{S}_w(\cdot)$ is mp-unbiased for $\boldsymbol{\beta}$ in the sense that $E_{mp}(\mathbf{S}_w(\boldsymbol{\beta})) = \mathbf{0}$. The role of the design weight w_k in the weighted estimating function $\mathbf{S}_w(\cdot)$ is to protect against selection bias caused by informative sampling. Informally speaking, informative sampling occurs when the model that holds for the sample units is not the same as the model that holds for the nonsample units. The design weights offer protection because, under certain conditions, it can be shown that $\widehat{\boldsymbol{\beta}}_{ws}$ is mp-consistent for $\boldsymbol{\beta}$ whether sampling is informative or not (e.g., Rubin-Bleuer and Schiopu-Kratina 2005). Note that the reference distribution for inference becomes the joint distribution induced by the model and the sampling design whereas it was only the sampling design in Sect. 2.

3.1 Bootstrap Variance Estimation

The objective is to estimate the total variance of $\widehat{\beta}_{ws}$:

$$\mathrm{var}_{mp}\left(\widehat{\beta}_{ws}\right) = E_m \mathrm{var}_p\left(\widehat{\beta}_{ws}\right) + \mathrm{var}_m E_p\left(\widehat{\beta}_{ws}\right). \tag{2}$$

The first term on the right-hand side of equation (2), $E_m \mathrm{var}_p\left(\widehat{\beta}_{ws}\right)$, is called the design variance whereas the second term, $\mathrm{var}_m E_p\left(\widehat{\beta}_{ws}\right)$, is called the model variance. Under certain conditions, the design variance is $O(n^{-1})$ whereas the model variance is $O(N^{-1})$. As a result, if the overall sampling fraction, n/N, is negligible, the model variance is negligible and the total variance can be estimated by finding an estimator of the design variance, as also noted by Pfeffermann (1993), Graubard and Korn (2002) and Binder and Roberts (2003).

An estimator of the design variance, $E_m \mathrm{var}_p\left(\widehat{\beta}_{ws}\right)$, is obtained by finding an estimator of $\mathrm{var}_p\left(\widehat{\beta}_{ws}\right)$. The bootstrap methodology described in Sect. 2 can be used to this end. The bootstrap design weights w_k^*, $k \in s$, can be used to compute $\widehat{\beta}_{w*s}$, the bootstrap replicate of $\widehat{\beta}_{ws}$, obtained through the estimating equation $S_{w*}\left(\widehat{\beta}_{w*s}\right) = 0$ with $S_{w*}(\beta) = \sum_{k \in s} w_k^* S_k(\beta)$. Beaumont and Patak (2012) showed through Taylor linearization that the bootstrap variance, $\mathrm{var}_*\left(\widehat{\beta}_{w*s}\right)$, is asymptotically equivalent to v_p, a p-consistent linearized estimator of the design variance $\mathrm{var}_p\left(\widehat{\beta}_{ws}\right)$. The bootstrap variance, $\mathrm{var}_*\left(\widehat{\beta}_{w*s}\right)$, is difficult to compute but can be approximated via Monte Carlo simulations by replicating w_k^*, $k \in s$, many times, as described at the end of Sect. 2.

The overall sampling fraction is not always negligible enough to ignore the model variance. Beaumont and Charest (2012) gave examples with real survey data illustrating this point (see also Graubard and Korn 2002). They also pointed out that the common practice of pretending with-replacement sampling, even though without-replacement sampling was actually used, may not be sufficient to capture the model variance. Below, we describe a methodology proposed by Beaumont and Charest (2012) to adjust the bootstrap design weights w_k^* so as to capture both the variability due to the sampling design and the model.

Let us define the adjusted bootstrap design weight, $w_k^{**} = w_k^* a_k$, where a_k is a bootstrap weight adjustment. The role of a_k is to capture the model variance so that w_k^{**} captures both the design and model variances. Let us also denote by $\widehat{\beta}_{w**s}$, the bootstrap replicate of $\widehat{\boldsymbol{\beta}}_{ws}$, obtained through the estimating equation $\mathbf{S}_{w**}\left(\widehat{\boldsymbol{\beta}}_{w**s}\right) = \mathbf{0}$ with $\mathbf{S}_{w**}\left(\boldsymbol{\beta}\right) = \sum_{k\in s} w_k^{**} \mathbf{S}_k\left(\boldsymbol{\beta}\right)$. The distribution of the adjustment a_k must be such that the bootstrap variance $\text{var}_*\left(\widehat{\boldsymbol{\beta}}_{w**s}\right)$ is mp-consistent for the total variance (2). Beaumont and Charest (2012) proposed to generate a_k, $k \in s$, independently with $E_*(a_k) = 1$ and

$$\text{var}_*\left(a_k\right) = \frac{w_k}{E_*\left[\left(w_k^*\right)^2\right]} = \frac{w_k}{\sigma_{kk} + w_k^2}. \tag{3}$$

The last equality in (3) comes from conditions (5) and (6). With this distribution of a_k, they showed through Taylor linearization that $\text{var}_*\left(\widehat{\boldsymbol{\beta}}_{w**s}\right) \approx v_p + v_m$, where v_p is mp-consistent for the design variance, $E_m\text{var}_p\left(\widehat{\boldsymbol{\beta}}_{ws}\right)$, and v_m is an mp-consistent estimator of the model variance, $\text{var}_m E_p\left(\widehat{\boldsymbol{\beta}}_{ws}\right)$, obtained through Taylor linearization.

Remark 1. If the Horvitz–Thompson variance estimator is used, $\sigma_{kk} = w_k(w_k - 1)$ and, from equation (3), we obtain: $\text{var}_*(a_k) = (2w_k - 1)^{-1}$. As a result, if w_k is large, $\text{var}_*(a_k)$ is close to 0 and the bootstrap weight adjustment a_k is close to 1. Therefore, if all the design weights w_k, $k \in s$, are large, the effect of the bootstrap adjustments is negligible. This makes sense since in that case the overall sampling fraction would be negligible and the model variance could be neglected.

Remark 2. The bootstrap variance, $\text{var}_*\left(\widehat{\boldsymbol{\beta}}_{w**s}\right)$, can again be approximated via Monte Carlo simulations by replicating $w_k^{**} = w_k^* a_k$, $k \in s$, many times. That is, both w_k^* and a_k must be replicated.

Remark 3. We have considered the case where the survey weights are given by $w_k = \pi_k^{-1}$. Sometimes, the survey weight w_k is calibrated to known population totals. As a result, the bootstrap weight w_k^* must also be calibrated. The above procedure for the generation of a_k continues to apply even if w_k and w_k^* are calibration weights. However, the expectation $E_*\left[\left(w_k^*\right)^2\right]$ in equation (3) does not have a closed form in general but can be approximated via Monte Carlo simulations by $B^{-1}\sum_{k\in s}(w_k^{*(b)})^2$.

3.2 Bootstrap Tests of Hypotheses

The survey analyst might be interested in testing a null hypothesis of the form $H_0 : \mathbf{H}\boldsymbol{\beta} = \mathbf{c}$ against the alternative hypothesis $H_1 : \mathbf{H}\boldsymbol{\beta} \neq \mathbf{c}$, where $\mathbf{H}$ is a matrix defining the hypotheses to be tested and $\mathbf{c}$ is a vector of constants, often set equal to 0. Let us consider weighted test statistics that have the following quadratic form:

$$t\left(\mathbf{w}_s; \mathbf{c}\right) = \left(\mathbf{H}\widehat{\boldsymbol{\beta}}_{ws} - \mathbf{c}\right)'\left(\widehat{\mathbf{A}}\left(\mathbf{w}_s\right)\right)^{-1}\left(\mathbf{H}\widehat{\boldsymbol{\beta}}_{ws} - \mathbf{c}\right),\tag{4}$$

where $\widehat{\mathbf{A}}\left(\mathbf{w}_s\right)$ is a scaling matrix and $\mathbf{w}_s$ is the vector containing the survey weights $w_k,\ k \in s$. There are many possible choices for the scaling matrix $\widehat{\mathbf{A}}\left(\mathbf{w}_s\right)$. A simple choice, recommended by Beaumont and Bocci (2009), is to compute $\widehat{\mathbf{A}}\left(\mathbf{w}_s\right)$ and $t(\mathbf{w}_s; \mathbf{c})$ using standard statistical software packages that allow users to weight each observation (e.g., SAS). The only restriction is that we assume that $\widehat{\mathbf{A}}\left(\mathbf{w}_s\right)$ and $t(\mathbf{w}_s; \mathbf{c})$ do not depend on the joint selection probabilities π_{kl}.

To obtain a valid test, it is necessary to approximate the unknown distribution of $t(\mathbf{w}_s; \mathbf{c})$ in equation (4) under the null hypothesis $H_0 : \mathbf{H}\boldsymbol{\beta} = \mathbf{c}$. We focus on bootstrap tests that approximate this distribution by using the bootstrap weights $w_k^{**},\ k \in s$. We denote by $\mathbf{w}_s^{**}$, the vector containing the bootstrap weights $w_k^{**},\ k \in s$. A natural choice would be to approximate the distribution of $t(\mathbf{w}_s; \mathbf{c})$ under the null hypothesis by the bootstrap distribution of $t(\mathbf{w}_s^{**}; \mathbf{c})$. It does not work because the bootstrap weights w_k^{**} do not reflect the null hypothesis since they have not been constructed using any null hypothesis in mind. Beaumont and Bocci (2009) showed that this issue can simply be solved by considering the bootstrap distribution of the statistic $t\left(\mathbf{w}_s^{**}; \mathbf{H}\widehat{\boldsymbol{\beta}}_{ws}\right)$ instead of $t(\mathbf{w}_s^{**}; \mathbf{c})$. They also showed in an empirical study that their bootstrap test performs as well as the method of Rao and Scott (1981) and better than Wald and Bonferroni tests.

4 Conclusions

We have shown that bootstrap variance estimation for estimators of finite population quantities can be achieved by generating many times bootstrap design weights w_k^*, $k \in s$. There is no need to generate many bootstrap samples. This simplifies bootstrap variance estimation for users as they are typically already familiar with the use of survey weights for obtaining estimates of population quantities. It suffices for the data producer to provide the bootstrap weights along with the survey data. Valid bootstrap variance estimation is possible for most sampling designs used in practice provided that the distribution used to generate the bootstrap weights satisfies the moment conditions (5) and (6) given in Sect. 2. These conditions are satisfied for most bootstrap methods found in the literature.

Each bootstrap method leads to a specific multivariate distribution of the bootstrap design weights. Ideally, we would like to use a distribution that leads to non-negative bootstrap design weights. Negative bootstrap design weights are not a problem per se if they are appropriately handled when computing bootstrap replicates of the sample estimates. But, many software packages cannot handle negative weights and, for this reason, it is usually more convenient for users if all the bootstrap design weights are non-negative. For some complex designs, such as the two-phase sampling design considered in Langlet et al. (2008), it seems difficult to find a multivariate distribution that ensures that the bootstrap design weights w_k^*, $k \in s$, are all non-negative while also satisfying the moment conditions (5) and (6). The multivariate lognormal distribution, which was proposed to us by Prof. Richard Lockhart (see also Beaumont and Patak 2012), is an option. However, it appeared through our empirical investigations that a multivariate lognormal distribution satisfying (5) and (6) does not always exist. Finding a multivariate distribution that satisfies (5) and (6) for general sampling designs while also ensuring that all the resulting bootstrap design weights are not negative is an issue that has not yet been resolved in the literature. An alternative method to deal with this issue of negative bootstrap design weights is the rescaling method of Beaumont and Patak (2012). The drawback of the method is that the bootstrap variance estimates must be multiplied by a constant factor that must be provided to the users.

Survey analysts may also be interested in estimating model parameters. For that case, we have described a simple method for adjusting the bootstrap weights so as to account for both the design and model variances. We have also shown how to use these adjusted bootstrap weights to perform bootstrap tests of hypotheses. This provides an alternative to the Rao–Scott method that is easy to implement as it simply involves replicating a standard test statistic of the form (4). Our method assumes that the observations are independent under the model. Developing suitable bootstrap weight adjustments when the independence assumption does not hold is an area that requires further research.

Acknowledgments I would like to thank the reviewer for the constructive comments that helped improve the overall quality of the paper.

References

Antal, E., Tillé, Y.: A direct bootstrap method for complex sampling designs from a finite population. J. Am. Stat. Assoc. **106**, 534–543 (2011)

Beaumont, J.-F., Bocci, C.: A practical bootstrap method for testing hypotheses from survey data. Surv. Methodol. **35**, 25–35 (2009)

Beaumont, J.-F., Charest, A.-S.: Bootstrap variance estimation with survey data when estimating model parameters. Comput. Stat. Data Anal. **56**, 4450–4461 (2012)

Beaumont, J.-F., Patak, Z.: On the generalized bootstrap for sample surveys with special attention to Poisson sampling. Int. Stat. Rev. **80**, 127–148 (2012)

Bertail, P., Combris, P.: Bootstrap généralisé d'un sondage. Annales d'économie et de statistique **46**, 49–83 (1997)

Binder, D.A.: On the variances of asymptotically normal estimates from complex surveys. Int. Stat. Rev. **51**, 279–292 (1983)

Binder, D.A., Roberts, G.R.: Design-based and model-based methods for estimating model parameters. In: Chambers, R.L., Skinner, C.J. (eds.) Analysis of Survey Data. Wiley, New York (2003)

Booth, J.G., Butler, R.W., Hall, P.: Bootstrap methods for finite populations. J. Am. Stat. Assoc. **89**, 1282–1289 (1994)

Chao, M.-T., Lo, S.-H.: A bootstrap method for finite population. Sankhya A. **47**, 399–405 (1985)

Chauvet, G.: Méthodes de bootstrap en population finie. PhD dissertation, Laboratoire de statistique d'enquêtes, CREST-ENSAI, Université de Rennes 2 (2007). http://tel.archives-ouvertes.fr/docs/00/26/76/89/PDF/thesechauvet.pdf

Chipperfield, J., Preston, J.: Efficient bootstrap for business surveys. Surv. Methodol. **33**, 167–172 (2007)

Demnati, A., Rao, J.N.K.: Linearization variance estimators for model parameters from complex survey data. Surv. Methodol. **36**, 193–201 (2010)

Fuller, W.A.: Sampling Statistics. Wiley, Hoboken (2009)

Graubard, B.I., Korn, E.L.: Inference for superpopulation parameters using sample surveys. Stat. Sci. **17**, 73–96 (2002)

Gross, S.: Median estimation in sample surveys. In: Proceedings of the Section on Survey Research Methods, American Statistical Association, Alexandria, pp. 181–184 (1980). http://www.amstat.org/Section/Srms/Proceedings/

Langlet, É., Beaumont, J.-F., Lavallée, P.: Bootstrap methods for two-phase sampling applicable to postcensal surveys. Paper presented at Statistics Canada's Advisory Committee on Statistical Methods, Ottawa, May 2008 (this paper is available upon request from the authors)

Pfeffermann, D.: The role of sampling weights when modeling survey data. Int. Stat. Rev. **61**, 317–337 (1993)

Rao, J.N.K., Scott, A.J.: The analysis of categorical data from complex sample surveys: chi-squared tests for goodness of fit and independence in two-way tables. J. Am. Stat. Assoc. **76**, 221–230 (1981)

Rao, J.N.K., Wu, C.F.J.: Resampling inference with complex survey data. J. Am. Stat. Assoc. **83**, 231–241 (1988)

Rao, J.N.K., Wu, C.F.J., Yue, K.: Some recent work on resampling methods for complex surveys. Surv. Methodol. **18**, 209–217 (1992)

Rao, J.N.K., Yung, W., Hidiroglou, M.: Estimating equations for the analysis of survey data using poststratification information. Sankhyā (Indian J. Stat. A) **64**, 364–378 (2002)

Rubin-Bleuer, S., Schiopu-Kratina, I.: On the two-phase framework for joint model and design-based inference. Ann. Stat. **33**, 2789–2810 (2005)

Sitter, R.R.: Comparing three bootstrap methods for survey data. Can. J. Stat. **20**, 135–154 (1992)

Empirical Likelihood Confidence Intervals: An Application to the EU-SILC Household Surveys

Yves G. Berger and Omar De La Riva Torres

Abstract Berger and De La Riva Torres (2012), proposed a proper empirical like-lihood approach which can be used to construct design-based confidence intervals. The proposed approach gives confidence intervals which may have better coverages than standard confidence intervals, which relies on normality, variance estimates and linearisation. The proposed approach does not rely on variance estimates, re-sampling or linearisation, even when the point estimator is not linear and does not have a normal distribution. It can be also used to construct confidence intervals of means, regressions coefficients, quantiles and poverty indicators. The proposed approach is less computational intensive than rescaled bootstrap (Rao and Wu 1988) which can be unstable and may not have the intended coverages (Berger and De La Riva Torres 2012). We apply the proposed approach to a measure of poverty based upon the European Union Statistics on Income and Living Conditions (EU-SILC) survey (Eurostat 2012). Confidence intervals of the persistent-risk-of-poverty indicator are estimated for the overall population and six sub-population domains determined by cross-classifying age groups and gender. This work was supported by consulting work for the Net-SILC2 project (Atkinson and Marlier 2010).

Keywords Design-based approach • Estimating equations • Hájek estimator • Horvitz–Thompson estimator • Stratification • Ultimate cluster approach • Unequal inclusion probabilities

Y.G. Berger (✉) • O. De La Riva Torres
University of Southampton, Southampton, SO17 1BJ, UK
e-mail: y.g.berger@soton.ac.uk

F. Mecatti et al. (eds.), *Contributions to Sampling Statistics*, Contributions to Statistics,
DOI 10.1007/978-3-319-05320-2_5,
© Springer International Publishing Switzerland 2014

1   Introduction

Every year since 2003, the *European Union Statistics on Income and Living Conditions* (EU-SILC) surveys collect information on income, poverty and social inclusion from approximately 300,000 households within the European Union, Switzerland, Norway, Iceland and Turkey (Eurostat 2012). The "Europe 2020" strategy for smart, sustainable and inclusive growth aims to lift at least 20 million people in the European Union from the risk of poverty and exclusion by 2020. Since the launch of this strategy, the importance of EU-SILC has grown even further: one of the five Europe 2020 headline targets is based on EU-SILC data. An European Commission Regulation (EC n28/2004 of 5th January 2004) stipulates that since the indicators based on EU-SILC are sample estimates, they should be reported along with standard errors estimates and confidence intervals, particularly if they are used for policy purposes.

In order to monitor poverty across the European Union, several poverty indicators are estimated. These indicators belong to the class of *Laeken* indicators which is composed of 18 complex indicators of poverty and inequality (Osier 2009). The *persistent-risk-of-poverty* indicator is one of the core EU-SILC longitudinal indicator. For a 4-year panel, this indicator is defined as the share of persons who are at risk of poverty at the fourth wave of the panel and at the two of the three preceding waves. The *at-the-risk-of-poverty* threshold is set at 60 % of the national median equivalised disposable income after social transfers (Eurostat 2012; Osier 2009). In this paper, we show how empirical likelihood confidence intervals can be computed for the persistent-risk-of-poverty indicator for sub-population domains. We use the 2009 EU-SILC user database.

Confidence intervals of EU-SILC poverty indicators are based upon normality and linearised variance estimates (Di Meglio et al. 2013). Alternative approach for confidence intervals are rescaled bootstrap (Rao and Wu 1988), pseudo empirical likelihood (Wu and Rao 2006) and empirical likelihood (Berger and De La Riva Torres 2012). Confidence intervals based upon the normality of the point estimator may not have the right coverages because the normality assumption may not hold when the variable of interest is skewed and/or contains many zeros. For example, the persistent risk of poverty is based on a dichotomous variables which contains many zeros. In this situation, empirical likelihood and pseudo empirical likelihood confidence intervals tend to have better coverages (Chen et al. 2003; Rao and Wu 2009). The proposed empirical likelihood approach is simpler to implement than the pseudo empirical likelihood and the bootstrap approaches. With bootstrap approaches, the tail coverages may be significantly different from their intended level (Berger and De La Riva Torres 2012). The bootstrap approaches are more computationally intensive, especially with calibration weights. They are also limited to negligible sampling fractions. For finite population, the asymptotic coverage of bootstrap confidence intervals has only been shown in few particular cases (e.g. Rao and Wu 1988), and is often only justified by simulation studies.

In Sect. 2, we introduced the proposed empirical likelihood approach. The maximum empirical likelihood estimator is defined in Sect. 2.1. The empirical likelihood confidence intervals is defined in Sect. 2.2. In Sect. 3, we show how the proposed approach can be implemented with EU-SILC. In Sect. 3.2, we give estimates of confidence intervals for several domains.

2 Empirical Likelihood Approach

Let U be a finite population of N units, where N is a fixed quantity which is not necessarily known. Suppose that the population parameter of interest θ_0 is the unique solution of the following estimating equation (Godambe 1960):

$$G(\theta) = 0, \quad \text{with } G(\theta) = \sum_{i \in U} g_i(\theta), \tag{1}$$

where $g_i(\theta)$ is a function of θ and of characteristics of the unit i. For example, for the persistent-risk-of-poverty $g_i(\theta) = y_i - \theta$, where $y_i = 1$ if the individual i is has a persistent-risk-of-poverty and $y_i = 0$ otherwise.

We may also be interested in sub-populations or domains estimates. The introduction of the domain membership function δ_{di} serves to divide the population U into D domains $\mathcal{U}_1, \dots, \mathcal{U}_d, \dots, \mathcal{U}_D$, where $\delta_{di} = 1$ if the i unit belongs to $\mathcal{U}_d$ and $\delta_{di} = 0$ otherwise (Särndal et al. 1992). If the parameter of interest is the persistent-risk-of-poverty for a domain $\mathcal{U}_d$, we use

$$g_i(\theta) = (y_i - \theta)\delta_{di}. \tag{2}$$

Suppose that we wish to estimate θ_0 from the data of a sample s of size n selected with a single stage unequal probabilities without replacement sampling design. We consider that the sample size n is a fixed (non-random) quantity. We adopt a non-parametric design-based approach, where the sampling distribution is specified by the sampling design and where the values of the variables of interest are fixed constants.

We propose to use the following *empirical log-likelihood function* (e.g. Berger and De La Riva Torres 2012; Owen 2001):

$$\ell(m) = \sum_{i=1}^{n} \log(m_i), \tag{3}$$

where m_i are unknown positive scale loads which will be estimated. Hartley and Rao (1969) showed that (3) is the correct empirical log-likelihood function under unequal probability sampling with replacement, where m_i/N can be interpretted as the probability to observe the i-th unit. Owen (2001, Chap. 6) showed that (3) is a

suitable empirical log-likelihood function when the units are selected independently
with a Poisson sampling design. Although under fixed size sampling designs, the
units are not selected independently, we propose to use the empirical log-likelihood
function (3) under fixed size sampling designs. The aim is to show that this empirical
log-likelihood function can be used for point estimation and to construct confidence
intervals (or to derive tests) under fixed size sampling designs.

The maximum likelihood estimators of m_i are the values $\hat{m}_i$ which maximise
$\ell(m)$ subject to the constraints $m_i \geq 0$ and

$$\sum_{i=1}^{n} m_i c_i = C, \tag{4}$$

where c_i is $Q \times 1$ vector associated with the i-th sampled unit and C is $Q \times 1$
vector. In Sect. 2.2, we give the expression of the c_i and C for stratified sampling
designs. Note that the c_i and C cannot be any vectors, as they must obey the
regularity conditions given by Berger and De La Riva Torres (2012). These are
standard conditions such as Liapounov type condition on the absolute moments
(e.g. Krewski and Rao 1981) and $\sqrt{n}$-consistency of the Horvitz and Thompson
(1952) estimator of C (e.g. Isaki and Fuller 1982). Most of the conditions proposed
by Berger and De La Riva Torres (2012) hold under more restrictive conditions
implicitly assumed by Deville and Särndal (1992) for calibration. We also need the
following condition on the inclusion probabilities $nN^{-1}\pi_i^{-1} = O(1)$ (e.g. Krewski
and Rao 1981, p. 1014). We consider that the c_i and C are such that there exists a
fixed vector t such that $t^{\top} c_i = \pi_i$ and $t^{\top} C = n$. This condition holds when the
design information (inclusion probabilities and stratification) are included in the c_i
(Berger and De La Riva Torres 2012).

Berger and De La Riva Torres (2012) showed that the solution of the maximisa-
tion of (3) subject to (4) is given by

$$\hat{m}_i = \left(\pi_i + \boldsymbol{\eta}^{\top} c_i\right)^{-1}, \tag{5}$$

where the quantity $\boldsymbol{\eta}$ is such that the constraint (4) holds. This quantity can be
computed using an iterative modified Newton–Raphson procedure (Polyak 1987).

2.1 Maximum Empirical Likelihood Estimator

In this section, we defined the maximum empirical likelihood estimate as the value
which minimises the empirical log-likelihood ratio function (6) defined below.

Let $\hat{m}_i$ be the values which maximise (3) subject to the constraints $m_i \geq 0$
and (4) for given c_i and C. Let $\ell(\hat{m}) = \sum_{i=1}^{n} \log(\hat{m}_i)$ be the maximum value of
the empirical log-likelihood function. Let $\hat{m}_i^*(\theta)$ be the values which maximise (3)
subject to the constraints $m_i \geq 0$ and $\sum_{i=1}^{n} m_i c_i^* = C^*$ with $c_i^* = (c_i^{\top}, g_i(\theta))^{\top}$

and $\boldsymbol{C}^* = (\boldsymbol{C}^\top, 0)^\top$. Let $\ell(\hat{m}^*, \theta) = \sum_{i=1}^n \log(\hat{m}_i^*(\theta))$. The *empirical log-likelihood ratio function* (or deviance or profile likelihood) is defined by the following function of θ:

$$\hat{r}(\theta) = 2 \{\ell(\hat{m}) - \ell(\hat{m}^*, \theta)\} \cdot \qquad (6)$$

The *maximum empirical likelihood estimate* $\hat{\theta}$ of θ_0 is defined by the value of θ which minimises the function $\hat{r}(\theta)$. As the minimum value of $\hat{r}(\theta)$ is zero, $\hat{\theta}$ is the solution of $\hat{r}(\theta) = 0$. It can be easily shown that this implies that $\hat{\theta}$ is the solution of the following estimating equation:

$$\hat{G}(\theta) = 0, \quad \text{with} \quad \hat{G}(\theta) = \sum_{i=1}^n \hat{m}_i \, g_i(\theta), \qquad (7)$$

where $\hat{m}_i$ is defined by (5). We assume that the $g_i(\theta)$ are such that $\hat{G}(\theta) = 0$ has a unique solution.

For example, when $c_i = \pi_i$ and $\boldsymbol{C} = n$, we have that $\boldsymbol{\eta} = 0$ and $m_i = \pi_i^{-1}$ and

$$\hat{G}(\theta) = \hat{G}(\theta)_\pi = \sum_{i=1}^n \frac{g_i(\theta)}{\pi_i}, \qquad (8)$$

is the Horvitz and Thompson (1952) estimator of the function (1). When θ_0 is the persistent-risk-of-poverty for a domain d, the $g_i(\theta)$ are given by (2) and

$$\hat{\theta} = \frac{\sum_{i=1}^n y_i \delta_{di} \pi_i^{-1}}{\sum_{i=1}^n \delta_{di} \pi_i^{-1}} \qquad (9)$$

which is the standard Hájek (1971) estimator of the persistent-risk-of-poverty for the domain $\mathcal{U}_d$.

2.2 Empirical Likelihood Confidence Intervals

The main advantage of the empirical likelihood approach is its capability of deriving non-parametric confidence intervals which do not depend on variance estimates. In this section, we propose to use the empirical log-likelihood ratio function (6) to derive empirical likelihood confidence intervals.

Suppose that the finite population U is stratified into H strata denoted by $U_1, \ldots, U_h, \ldots, U_H$, where $\cup_{h=1}^H U_h = U$. Suppose that a sample s_h of fixed size n_h is selected with replacement with unequal probabilities from U_h. We assume that the number of strata H is asymptotically bounded.

The empirical likelihood estimator is still the solution of (7) where $\hat{m}_i$ are the values which maximise (3) under a set of constraints (4) with $c_i = z_i$ and $C = n$, where z_i are the values of the design (or stratification) variables defined by

$$z_i = (z_{i1}, \ldots, z_{iH})^\top \quad \text{and where} \quad n = (n_1, \ldots, n_H)^\top \tag{10}$$

denotes the vector of the strata sample sizes, with $z_{ih} = \pi_i$ when $i \in U_h$ and $z_{ih} = 0$ otherwise. It can be shown that $\hat{m}_i = \pi_i^{-1}$.

Berger and De La Riva Torres (2012) showed that

$$\hat{r}(\theta_0) = \hat{G}(\theta_0)_\pi^2 \, \widehat{\text{var}}_{\text{pps}}[\hat{G}(\theta_0)_\pi]^{-1} + o_p(1), \tag{11}$$

where $\hat{r}(\theta)$ is defined by (6) and the random variable $\widehat{\text{var}}_{\text{pps}}[\hat{G}(\theta_0)_\pi]$ is now the following stratified variance pps estimator

$$\widehat{\text{var}}_{\text{st}}[\hat{G}(\theta_0)_\pi] = \sum_{h=1}^{H} \left[\sum_{i \in s_h} \left(\frac{g_i(\theta_0)}{\pi_i} - \frac{\hat{G}_{\pi;h}(\theta_0)}{n_h} \right)^2 \right], \tag{12}$$

where $\hat{G}_{\pi;h}(\theta_0) = \sum_{i \in s_h} g_i(\theta_0)\pi_i^{-1}$. This variance estimator is consistent when the number of strata is bounded and when the sampling fraction is negligible. Thus, the random variable $\hat{r}(\theta_0)$ follows asymptotically a chi-squared distribution with one degree of freedom because $\hat{G}(\theta_0)_\pi$ is an Horvitz and Thompson (1952) estimator which follows a normal distribution asymptotically (Berger 1998; Hájek 1964).

Note that the normality of $\hat{G}(\theta_0)_\pi$ implies the asymptotic normality of $\hat{\theta}$ under additional regularity conditions. The empirical likelihood confidence interval relies on the normality of $\hat{G}(\theta_0)_\pi$ and not on the normality of $\hat{\theta}$. For non-linear estimator, such as quantiles, the distribution of $\hat{\theta}$ may not be approximately normal. However, $\hat{G}(\theta_0)_\pi$ is more likely to be normal, because $\hat{G}(\theta_0)_\pi$ is an Horvitz and Thompson (1952) estimator.

As $\hat{r}(\theta_0)$ follows a chi-squared distribution, the α level empirical likelihood confidence interval (e.g. Hudson 1971; Wilks 1938) for the population parameter θ_0 is given by

$$\{\theta \, : \, \hat{r}(\theta) \leq \chi_1^2(\alpha)\} = \left[\min\{\theta | \, \hat{r}(\theta) \leq \chi_1^2(\alpha)\} \, ; \, \max\{\theta | \, \hat{r}(\theta) \leq \chi_1^2(\alpha)\} \right], \tag{13}$$

where $\chi_1^2(\alpha)$ is the upper α-quantile of the chi-squared distribution with one degree of freedom. The interval (13) is indeed a confidence interval because it includes all the values θ which as such that the hypothesis $\theta_0 = \theta$ is not rejected using the test statistics $\hat{r}(\theta)$. Note that $\hat{r}(\theta)$ is a convex non-symmetric function with a minimum when θ is the maximum empirical likelihood estimator. This interval can be found using a bijection search method (e.g. Wu 2005). This involves calculating $\hat{r}(\theta)$ for several values of θ. Note that the bias of $\hat{\theta}$ does not affect the coverage of (13). Berger and De La Riva Torres (2012) showed how this approach can be used to accommodate large sampling fractions and non-response. Calibration constraints can also be taken into account by including the auxiliary variables within the c_i.

The pseudo empirical likelihood approach proposed by Chen and Sitter (1999) is not entirely satisfactory, because (a) its empirical likelihood function is not a standard one (b) and its empirical log-likelihood ratio function does not converge to a chi-squared distribution (Wu and Rao 2006). For confidence interval, the pseudo empirical log-likelihood ratio function needs to be adjusted by a ratio of variances (the design effect) which needs to be estimated. This has two major drawbacks: it may affect the coverage and it limits the range of the parameters it can be applied to. The proposed approach is a proper empirical likelihood approach which does not rely on design effects, and is simpler to implement than the pseudo empirical likelihood approach. It can be used for a wider class of parameters of interest.

3   Confidence Intervals for Persistent-Risks-of-Poverty

When $g_i(\theta)$ is given by (2), the interval (13) gives a confidence interval for the persistent-risk-of-poverty, if we have a single stage stratified sampling design. However, most of the EU-SILC surveys are multi-stage cluster surveys. Therefore, the approach described in Sect. 2.2 cannot be directly implemented. We proposed to use an ultimate cluster approach which consists in treating the primary sampling unit (PSU) totals estimates of $g_i(\theta)$ and the PSU as sampling units (Hansen et al. 1953). Within each PSU, sub-sampling of any complexity may be involved. It is common to have small sampling fractions with social surveys. Thus, the variance estimator (12) is approximately unbiased and $\hat{r}(\theta_0)$ follows asymptotically a chi-squared distribution with one degree of freedom. The at-the-risk-of-poverty rate estimators contains sampling errors due to the estimation of the poverty thresholds. For simplicity, we assume that the poverty threshold is fixed which ensures conservative confidence intervals (Berger and Skinner 2003).

Calibration and stratification variables were not available. We used the Nomenclature of Territorial Units for Statistics (NUTS2) geographical region as a proxy for stratification and the effect of calibration was ignored. If the calibration variables were available, we suggest including them within the c_i. This would give confidence intervals which take into account of calibration (Berger and De La Riva Torres 2012).

The confidence intervals for the persistent-risk-of-poverty rate are estimated for the overall population (Sect. 3.1) and domains defined by gender and age groups (Sect. 3.2). We adopted the approach described in Sect. 2 to construct confidence intervals, treating the sample of PSU's as a single stage sample. Standard confidence intervals and rescaled bootstrap confidence intervals (Rao and Wu 1988; Rao et al.

1992) are also computed. The former assumes normal distributed data for its variance estimation, and the latter is based on re-sampling techniques.

3.1 Confidence Intervals for Population Estimators

In Table 1, we have the confidence intervals for the population estimators. The empirical likelihood confidence intervals are consistent with the standard confidence intervals. The sampling distribution must be approximately normal, as we do not observe large differences between these intervals. Nevertheless, the bounds of the empirical likelihood intervals are always larger than the bounds of the standard intervals and the bootstrap intervals tend to be located between empirical likelihood intervals and standard interval. These differences is particularly pronounced for Latvia. Differences observed are due to the skewness of the sampling distribution which is caused by the skewness of the sampling distribution of the estimator of the persistent-risk-of-poverty.

3.2 Confidence Intervals for Domains Estimators

We consider several domains of interest by cross-classifying gender and three age groups: (a) 16-year old to 25-year old; (b) 25-year old to 44-year old and (c) older than 44-year old. This produces six domains. The differences observed between confidence intervals are due to the skewness of the sampling distribution. The differences observed between the standard, the bootstrap and the empirical likelihood confidence intervals are due to the fact that domain point estimators may not be normal. The domain size affects the length of the confidence intervals, but not the asymmetry of the empirical likelihood confidence intervals. The asymmetry is the result of the sampling distribution which is affected by the inclusion probabilities, the sample size and the skewness of the variable of interest. The confidence intervals vary from countries to countries because of the differences in income distribution and sampling design.

Table 2 gives the results for men and women. These results show the same pattern as in Sect. 3.1. Furthermore, Ireland, Lithuania and Latvia showed the most noticeable differences, particularly for women. For the Netherlands and Poland, men's confidence intervals are noticeably different.

The confidence intervals of the individuals younger than 25 years old are given in Table 3. Latvia exhibited the highest differences in the upper bounds. The standard confidence intervals for Iceland is an example of poor confidence interval, as the lower bound is negative. Note that for Poland, the standard confidence interval is wider than confidence intervals based on empirical likelihood and bootstrap.

The results for the age group 25–44 years old are given in Table 4. Note that the bounds of the standard intervals for men are negative for Austria, Denmark, Ireland

Table 1 Estimates of persistent-risk-of-poverty rate for 2009

Country	Rate (%)	Lower (%)			Upper (%)		
Austria	5.9	4.6	(4.4)	[4.5]	7.7	(7.5)	[7.6]
Belgium	8.9	7.1	(6.9)	[7.0]	11.3	(11.0)	[11.1]
Bulgaria	10.7	8.2	(7.9)	[8.0]	13.9	(13.4)	[13.6]
Cyprus	11.0	9.1	(8.9)	[9.1]	13.1	(13.0)	[12.9]
Czech Republic	3.6	2.5	(2.3)	[2.4]	5.1	(4.8)	[5.0]
Denmark	6.3	4.6	(4.4)	[4.5]	8.4	(8.2)	[8.2]
Estonia	13.0	11.1	(10.9)	[11.1]	15.1	(15.0)	[15.1]
Spain	11.1	9.5	(9.4)	[9.4]	12.9	(12.8)	[12.8]
Finland	6.9	5.5	(5.4)	[5.5]	8.6	(8.4)	[8.3]
France	6.9	6.0	(6.0)	[5.9]	7.9	(7.8)	[8.0]
Greece	14.5	11.8	(11.8)	[11.5]	17.8	(17.2)	[17.6]
Hungary	8.3	6.6	(6.5)	[6.5]	10.4	(10.1)	[10.3]
Ireland	6.3	4.3	(3.7)	[4.1]	9.5	(8.9)	[8.9]
Iceland	4.2	2.4	(2.0)	[2.0]	6.9	(6.4)	[6.7]
Italy	13.4	11.2	(11.3)	[10.9]	16.0	(15.5)	[16.1]
Lithuania	11.7	9.2	(8.9)	[9.1]	15.2	(14.5)	[15.0]
Luxembourg	8.8	6.9	(6.7)	[6.9]	11.5	(11.0)	[11.2]
Latvia	17.7	14.0	(13.5)	[13.6]	23.9	(22.0)	[23.1]
Malta	6.2	4.6	(4.4)	[4.5]	8.2	(8.0)	[8.1]
The Netherlands	6.4	4.1	(3.7)	[3.7]	9.8	(9.0)	[9.4]
Norway	5.4	4.2	(4.0)	[4.1]	6.9	(6.7)	[6.8]
Poland	10.2	7.5	(8.6)	[7.4]	13.4	(11.7)	[13.4]
Portugal	10.0	7.8	(7.6)	[7.7]	12.7	(12.4)	[12.5]
Sweden	5.7	4.4	(4.2)	[4.2]	7.2	(7.1)	[7.0]
Slovakia	5.0	3.5	(3.3)	[3.4]	7.1	(6.7)	[6.9]
United Kingdom	8.4	6.7	(6.5)	[6.5]	10.5	(10.2)	[10.4]

95 % empirical likelihood confidence intervals. The standard confidence intervals are given in round brackets. Rescaled bootstrap confidence intervals are given in squared brackets

and Malta. The bootstrap bounds and the empirical likelihood bounds are larger than the bounds of the standard intervals. These differences are more pronounced for Austria, Malta, Denmark, the Netherlands, Estonia, Latvia and Greece. This is due to the skewness of the sampling distribution. For Poland, the standard confidence interval is smaller than the other confidence intervals. The lower bound of standard confidence interval for the women in the Netherlands is negative. Ireland and Lithuania give the largest differences between the three confidence intervals. The bootstrap bounds tend to be smaller than bounds of the empirical likelihood intervals. This is probably due to the fact that with the bootstrap the lower tail coverage tends to be too low and the upper tail coverage tends to be too large. Berger and De La Riva Torres (2012) observed this fact through a series of simulation.

In Table 5, we consider the age group above 44-year old. Standard confidence interval for Iceland has negative lower bound. Finland, Denmark, Greece and Latvia have the most noticeable difference, particularly in the upper bound for the men.

Table 2 Estimates of persistent-risk-of-poverty rate for 2009 by gender

Country	Rate (%)	Lower (%)	Upper (%)	Country	Rate (%)	Lower (%)	Upper (%)
Austria				Iceland			
Men	4.6	3.1 (2.9) [3.0]	6.9 (6.4) [6.6]	Men	3.3	1.6 (1.2) [1.2]	6.1 (5.4) [5.5]
Women	7.1	5.5 (5.2) [5.3]	9.3 (9.0) [8.9]	Women	5.1	2.7 (2.2) [2.4]	8.6 (7.9) [8.1]
Belgium				Italy			
Men	7.8	5.9 (5.7) [5.8]	10.1 (9.8) [9.8]	Men	12.3	10.0 (10.0) [9.9]	15.1 (14.5) [14.9]
Women	10	7.8 (7.5) [7.5]	13.1 (12.5) [12.9]	Women	14.5	12.1 (12.3) [11.9]	17.1 (16.6) [17.2]
Bulgaria				Lithuania			
Men	9.7	7.1 (6.8) [6.9]	13.0 (12.6) [12.5]	Men	9.4	6.8 (6.5) [6.7]	12.9 (12.3) [12.6]
Women	11.6	8.9 (8.6) [8.8]	15.1 (14.5) [14.9]	Women	13.7	10.8 (10.5) [10.6]	17.9 (16.9) [17.6]
Cyprus				Luxembourg			
Men	8.3	6.6 (6.4) [6.6]	10.3 (10.2) [10.2]	Men	7.8	5.8 (5.5) [5.6]	10.6 (10.0) [10.0]
Women	13.5	11.2 (11.0) [11.2]	16.1 (15.9) [15.8]	Women	9.9	7.6 (7.3) [7.4]	12.9 (12.3) [12.6]
Czech Republic				Latvia			
Men	3.1	2.0 (1.6) [1.8]	5.3 (4.6) [4.9]	Men	15.1	11.0 (10.6) [10.5]	21.3 (19.6) [20.4]
Women	3.9	2.9 (2.8) [2.8]	5.5 (5.1) [5.3]	Women	20.0	15.9 (15.3) [15.6]	26.9 (24.6) [25.8]
Denmark				Malta			
Men	5.9	3.9 (3.6) [3.5]	8.6 (8.1) [8.3]	Men	5.5	3.8 (3.5) [3.6]	8.1 (7.6) [7.7]
Women	6.7	4.6 (4.3) [4.3]	9.5 (9.0) [9.0]	Women	6.9	5.1 (4.9) [5.0]	9.2 (8.9) [9.0]
Estonia				Netherlands			
Men	11.6	9.3 (9.0) [9.3]	14.5 (14.1) [13.9]	Men	7.4	4.4 (3.8) [4.0]	12.4 (11.0) [11.7]
Women	14.1	11.9 (11.8) [11.8]	16.6 (16.5) [16.6]	Women	5.4	3.2 (2.8) [3.0]	8.7 (8.0) [8.4]
Spain				Norway			
Men	10.5	8.8 (8.7) [8.6]	12.5 (12.3) [12.3]	Men	3.8	2.7 (2.5) [2.7]	5.3 (5.0) [5.2]
Women	12.6	10.9 (10.8) [10.7]	14.6 (14.5) [14.5]	Women	6.9	4.9 (4.7) [4.8]	9.6 (9.0) [9.3]
Finland				Poland			
Men	5.5	4.1 (3.9) [4.1]	7.5 (7.2) [7.0]	Men	10.3	7.5 (8.6) [7.5]	13.7 (11.9) [13.4]
Women	8.2	6.4 (6.2) [6.1]	10.5 (10.3) [10.3]	Women	10.1	7.5 (8.5) [7.3]	13.3 (11.6) [13.1]

Country							
France							
Men	6.2	5.3	(5.2)	[5.1]	7.3	(7.2)	[7.3]
Women	7.6	6.6	(6.6)	[6.5]	8.7	(8.6)	[8.7]
Greece							
Men	13.9	10.9	(10.8)	[10.7]	18.1	(17.1)	[17.5]
Women	15.1	12.2	(12.2)	[12.1]	18.4	(17.9)	[18.1]
Hungary							
Men	8.8	7.0	(6.8)	[6.7]	11.2	(10.9)	[10.8]
Women	7.8	6.2	(6.0)	[6.0]	9.9	(9.6)	[9.7]
Ireland							
Men	5.9	3.8	(3.3)	[3.6]	8.8	(8.4)	[8.4]
Women	6.8	4.5	(3.7)	[4.2]	10.7	(9.8)	[9.8]
Portugal							
Men	9.4	7.1	(6.8)	[7.0]	12.4	(12.0)	[12.3]
Women	10.5	8.2	(7.9)	[8.0]	13.6	(13.1)	[13.2]
Sweden							
Men	4.5	3.3	(3.1)	[3.2]	6.1	(5.9)	[5.9]
Women	6.8	5.1	(4.9)	[5.0]	8.8	(8.6)	[8.7]
Slovakia							
Men	4.8	3.0	(2.6)	[2.7]	7.5	(6.9)	[6.9]
Women	5.3	3.8	(3.6)	[3.7]	7.1	(6.9)	[6.9]
United Kingdom							
Men	8.0	6.2	(6.0)	[6.1]	10.2	(10.0)	[9.9]
Women	8.7	6.8	(6.6)	[6.7]	11.2	(10.8)	[11.1]

95 % empirical likelihood confidence intervals. The standard confidence intervals are given in round brackets. Rescaled bootstrap confidence intervals are given in squared brackets

Table 3 Estimates of persistent-risk-of-poverty rate for 2009 by gender and age group <25 years old

Country	Rate (%)	Lower (%)	Upper (%)
Austria			
Men	2.1	0.7 (0.2) [0.3]	5.0 (4.2) [4.4]
Women	3.9	1.4 (0.5) [0.8]	8.7 (7.3) [7.9]
Belgium			
Men	8.9	5.6 (4.9) [5.2]	13.6 (12.8) [12.8]
Women	9.2	4.7 (3.5) [3.9]	17.6 (15.0) [15.7]
Bulgaria			
Men	15.6	10.1 (9.8) [9.3]	23.1 (21.5) [22.5]
Women	11.2	6.5 (5.7) [5.9]	18.2 (16.7) [17.6]
Cyprus			
Men	4.5	2.4 (1.8) [2.1]	7.8 (7.2) [7.3]
Women	7.3	4.2 (3.6) [3.8]	11.9 (11.0) [11.5]
Czech Republic			
Men	6.2	3.3 (2.4) [2.9]	12.3 (10.1) [10.9]
Women	4.8	2.6 (1.9) [2.1]	9.1 (7.8) [8.2]
Denmark			
Men	3.0	1.1 (0.4) [0.6]	6.7 (5.6) [6.0]
Women	3.4	1.1 (0.0) [0.3]	8.4 (6.8) [7.0]
Estonia			
Men	13.9	9.6 (9.0) [9.0]	20.4 (18.8) [19.6]
Women	10.4	7.2 (6.7) [6.8]	15.0 (14.2) [14.5]
Spain			
Men	12.8	9.7 (9.5) [9.7]	16.8 (16.2) [16.9]
Women	15.6	11.9 (11.7) [11.7]	20.3 (19.5) [19.6]
Finland			
Men	3.4	1.9 (1.5) [1.7]	5.9 (5.2) [5.4]
Women	5.4	3.2 (2.6) [3.2]	8.9 (8.2) [8.3]
Iceland			
Men	3.3	0.9 (−0.2) [0.4]	8.4 (6.8) [7.2]
Women	6.7	3.0 (2.1) [2.3]	12.7 (11.3) [12.0]
Italy			
Men	16.5	12.3 (12.1) [12.0]	22.1 (21.0) [22.1]
Women	17.9	13.4 (13.9) [13.2]	23.2 (21.8) [23.2]
Lithuania			
Men	11.1	6.0 (5.1) [5.3]	19.4 (17.1) [18.6]
Women	15.4	9.2 (8.5) [8.2]	25.2 (22.4) [24.1]
Luxembourg			
Men	13.2	9.1 (8.4) [8.8]	19.6 (18.0) [19.0]
Women	15.3	10.5 (9.9) [10.0]	22.5 (20.6) [21.4]
Latvia			
Men	17.6	9.8 (8.7) [7.9]	33.4 (26.4) [31.1]
Women	21.2	11.9 (9.8) [10.3]	42.5 (32.6) [36.6]
Malta			
Men	6.0	3.3 (2.6) [3.0]	10.3 (9.4) [9.6]
Women	7.4	4.3 (3.7) [3.9]	12.0 (11.2) [11.5]
Netherlands			
Men	12.5	5.4 (3.9) [4.3]	25.5 (21.0) [23.8]
Women	5.4	1.8 (0.5) [1.1]	12.6 (10.4) [11.0]
Norway			
Men	5.8	3.4 (2.9) [3.1]	9.8 (8.7) [9.0]
Women	7.9	4.9 (4.5) [4.7]	12.6 (11.4) [12.0]
Poland			
Men	14.7	10.5 (11.8) [10.2]	20.2 (17.6) [20.0]
Women	15.5	11.1 (12.4) [10.5]	21.6 (18.6) [21.0]

France							
Men	8.3	6.6	(6.5)	[6.6]	10.5	(10.2)	[10.3]
Women	10.1	8.0	(7.9)	[7.9]	12.6	(12.2)	[12.4]
Greece							
Men	17.0	11.8	(11.6)	[11.2]	23.6	(22.3)	[22.8]
Women	14.9	10.6	(10.4)	[10.1]	20.8	(19.5)	[20.2]
Hungary							
Men	16.6	12.2	(12.0)	[12.0]	22.5	(21.2)	[22.1]
Women	13.1	9.0	(8.6)	[8.6]	18.8	(17.5)	[18.1]
Ireland							
Men	9.4	4.7	(3.4)	[3.8]	17.0	(15.3)	[16.1]
Women	5.1	2.4	(1.1)	[2.0]	10.6	(9.0)	[9.1]
Portugal							
Men	10.2	6.0	(5.2)	[5.5]	16.6	(15.2)	[15.5]
Women	9.7	5.7	(4.9)	[5.3]	15.8	(14.5)	[15.1]
Sweden							
Men	5.1	3.0	(2.6)	[2.6]	8.1	(7.6)	[7.9]
Women	6.0	3.3	(2.7)	[3.0]	10.3	(9.2)	[9.6]
Slovakia							
Men	7.0	3.7	(2.8)	[3.2]	13.1	(11.2)	[12.3]
Women	6.1	3.4	(2.9)	[3.1]	10.1	(9.3)	[9.8]
United Kingdom							
Men	9.8	6.3	(5.9)	[6.3]	14.6	(13.8)	[14.2]
Women	9.9	5.6	(4.8)	[5.1]	17.2	(15.1)	[16.4]

95 % empirical likelihood confidence intervals. The standard confidence intervals are given in round brackets. Rescaled bootstrap confidence intervals are given in squared brackets

Table 4 Estimates of persistent-risk-of-poverty rate for 2009 by gender and age group between 25- and 44-year old

Country	Rate (%)	Lower (%)	Upper (%)	Country	Rate (%)	Lower (%)	Upper (%)
Austria				Iceland			
Men	2.1	0.5 (−0.5) [0.1]	6.5 (4.8) [5.3]	Men	3.6	1.4 (0.6) [0.9]	7.5 (6.5) [7.3]
Women	4.7	2.5 (1.9) [2.0]	8.4 (7.5) [7.7]	Women	6.4	2.8 (1.8) [2.0]	12.7 (11.1) [11.7]
Belgium				Italy			
Men	5.6	3.0 (2.4) [2.6]	9.9 (8.9) [9.4]	Men	9.5	7.1 (7.1) [6.9]	12.8 (12.0) [12.4]
Women	8.6	5.5 (5.1) [5.1]	12.8 (12.1) [12.1]	Women	12.5	9.7 (9.9) [9.4]	15.9 (15.1) [15.6]
Bulgaria				Lithuania			
Men	6.6	3.7 (3.1) [3.3]	11.0 (10.1) [10.4]	Men	8.0	4.7 (4.1) [4.3]	12.9 (12.0) [12.9]
Women	8.9	5.3 (4.6) [4.9]	14.3 (13.1) [13.8]	Women	9.8	5.7 (5.0) [5.5]	16.2 (14.6) [15.0]
Cyprus				Luxembourg			
Men	2.4	1.0 (0.6) [0.7]	4.9 (4.3) [4.4]	Men	6.4	4.4 (4.0) [4.2]	9.5 (8.9) [9.1]
Women	4.4	2.4 (2.0) [2.1]	7.3 (6.8) [7.1]	Women	10.6	7.7 (7.3) [7.3]	14.6 (13.8) [13.9]
Czech Republic				Latvia			
Men	1.5	0.7 (0.3) [0.4]	3.3 (2.8) [3.0]	Men	10.3	6.1 (5.0) [5.4]	17.4 (15.6) [15.3]
Women	3.5	2.1 (1.8) [1.9]	5.8 (5.3) [5.4]	Women	9.4	6.0 (5.6) [5.9]	14.3 (13.2) [13.5]
Denmark				Malta			
Men	3.5	1.1 (−0.1) [0.7]	8.9 (7.0) [7.8]	Men	2.9	1.0 (−0.1) [0.6]	7.8 (5.9) [6.1]
Women	4.1	1.6 (0.8) [1.2]	8.5 (7.3) [7.6]	Women	3.1	1.5 (1.0) [1.2]	5.8 (5.2) [5.3]
Estonia				Netherlands			
Men	7.4	4.1 (2.9) [3.5]	14.7 (12.0) [13.1]	Men	5.2	1.9 (0.7) [1.3]	11.7 (9.8) [10.2]
Women	7.3	5.0 (4.5) [4.9]	10.7 (10.2) [10.2]	Women	3.6	1.0 (−0.2) [0.4]	9.3 (7.4) [8.1]
Spain				Norway			
Men	7.7	5.8 (5.5) [5.5]	10.2 (9.9) [10.1]	Men	4.1	2.4 (2.0) [2.1]	6.9 (6.2) [6.3]
Women	9.0	7.0 (6.8) [7.0]	11.5 (11.2) [11.4]	Women	3.4	2.0 (1.7) [1.8]	5.5 (5.0) [5.0]
Finland				Poland			
Men	2.9	1.5 (1.2) [1.3]	5.1 (4.6) [4.9]	Men	8.6	5.9 (6.3) [5.3]	12.5 (10.8) [12.1]
Women	2.4	1.3 (0.9) [1.0]	4.2 (3.8) [3.9]	Women	9.4	6.9 (7.9) [6.7]	12.5 (11.0) [12.6]

France						
Men	4.5	3.3	(3.2)	[3.2]	6.0	(5.8) [6.0]
Women	7.3	5.8	(5.6)	[5.7]	9.2	(8.9) [9.0]
Greece						
Men	11.3	7.5	(6.7)	[7.0]	18.3	(16.0) [17.0]
Women	16.8	12.4	(12.4)	[12.2]	22.3	(21.2) [21.8]
Hungary						
Men	7.0	5.0	(4.7)	[4.9]	9.7	(9.3) [9.5]
Women	8.3	6.0	(5.8)	[5.7]	11.2	(10.8) [10.7]
Ireland						
Men	0.5	0.1	(−0.3)	[0.0]	1.8	(1.3) [1.6]
Women	5.3	2.0	(0.3)	[1.3]	12.8	(10.3) [10.7]
Portugal						
Men	7.1	4.2	(3.7)	[3.7]	11.3	(10.6) [10.9]
Women	7.0	3.9	(3.2)	[3.4]	12.0	(10.8) [11.3]
Sweden						
Men	3.8	2.1	(1.7)	[2.0]	6.4	(6.0) [6.2]
Women	4.2	2.4	(2.1)	[2.1]	6.8	(6.4) [6.7]
Slovakia						
Men	3.0	1.5	(1.2)	[1.3]	5.1	(4.7) [4.8]
Women	3.8	2.1	(1.7)	[1.9]	6.5	(5.9) [6.0]
United Kingdom						
Men	5.2	2.6	(1.8)	[2.2]	9.9	(8.6) [8.9]
Women	7.3	4.5	(4.1)	[4.3]	11.2	(10.5) [10.8]

95 % empirical likelihood confidence intervals. The standard confidence intervals are given in round brackets. Rescaled bootstrap confidence intervals are given in squared brackets

Table 5 Estimates of persistent-risk-of-poverty rate for 2009 by gender and age group >44 years old

Country	Rate (%)	Lower (%)	Upper (%)	Country	Rate (%)	Lower (%)	Upper (%)
Austria				Iceland			
Men	7.3	5.0 (4.6) [4.6]	10.8 (10.1) [10.6]	Men	3.1	1.0 (−0.2) [0.5]	8.4 (6.5) [7.0]
Women	9.5	7.4 (7.2) [7.3]	12.3 (11.9) [12.3]	Women	3.2	1.0 (0.1) [0.6]	7.6 (6.2) [6.3]
Belgium				Italy			
Men	8.4	6.0 (5.8) [5.7]	11.4 (11.0) [11.1]	Men	11.8	9.7 (9.7) [9.5]	14.3 (13.9) [14.1]
Women	11.1	8.4 (8.2) [8.2]	14.4 (13.9) [14.4]	Women	14.1	12.1 (11.9) [12.1]	16.4 (16.3) [16.3]
Bulgaria				Lithuania			
Men	8.8	6.5 (6.3) [6.5]	11.6 (11.3) [11.6]	Men	9.1	6.6 (6.4) [6.6]	12.2 (11.7) [11.8]
Women	13.0	10.2 (9.9) [9.9]	16.4 (16.1) [16.0]	Women	14.7	12.1 (12.0) [12.0]	17.8 (17.4) [17.6]
Cyprus				Luxembourg			
Men	14.4	11.4 (11.3) [11.3]	17.9 (17.6) [17.7]	Men	5.3	3.5 (3.1) [3.1]	8.4 (7.6) [7.9]
Women	21.8	18.3 (18.2) [18.2]	25.8 (25.4) [25.5]	Women	6.5	4.5 (4.1) [4.2]	9.6 (8.9) [8.9]
Czech Republic				Latvia			
Men	2.4	1.5 (1.3) [1.4]	3.8 (3.5) [3.6]	Men	17.1	12.9 (12.4) [12.7]	23.0 (21.9) [22.1]
Women	3.8	2.9 (2.9) [2.8]	4.9 (4.7) [4.8]	Women	25.0	20.6 (20.5) [20.3]	31.2 (29.4) [30.1]
Denmark				Malta			
Men	8.8	5.7 (5.2) [5.2]	13.5 (12.5) [13.1]	Men	6.7	4.3 (3.6) [3.9]	11.3 (9.8) [10.2]
Women	9.4	6.3 (6.0) [6.1]	13.7 (12.9) [12.9]	Women	8.8	6.3 (6.1) [6.1]	11.7 (11.4) [11.4]
Estonia				Netherlands			
Men	12.8	9.8 (9.6) [9.9]	16.5 (16.0) [16.3]	Men	5.5	3.3 (2.9) [3.0]	8.8 (8.2) [8.5]
Women	19.3	15.9 (15.8) [15.7]	23.3 (22.7) [23.2]	Women	6.3	3.7 (3.3) [3.4]	10.2 (9.3) [9.6]
Spain				Norway			
Men	11.2	9.3 (9.2) [9.3]	13.4 (13.2) [13.3]	Men	2.4	1.2 (0.8) [0.9]	4.6 (4.0) [4.3]
Women	13.4	11.2 (11.1) [11.0]	15.9 (15.7) [15.8]	Women	8.1	4.9 (4.4) [4.3]	12.9 (11.8) [12.3]
Finland				Poland			
Men	8.9	6.1 (5.7) [5.8]	13.2 (12.0) [12.6]	Men	8.5	6.3 (7.0) [6.1]	11.3 (10.1) [11.2]
Women	13.2	9.7 (9.5) [9.2]	17.5 (16.8) [17.2]	Women	7.7	5.7 (6.4) [5.4]	10.2 (9.0) [9.9]

France						
Men	5.8	4.7	(4.7)	[4.7]	7.1	(6.9) [7.0]
Women	6.5	5.6	(5.5)	[5.5]	7.6	(7.5) [7.7]
Greece						
Men	14.1	11.0	(10.8)	[10.7]	18.4	(17.4) [17.9]
Women	14.0	11.1	(11.0)	[11.0]	17.8	(17.0) [17.5]
Hungary						
Men	4.8	3.6	(3.4)	[3.3]	6.4	(6.2) [6.2]
Women	5.0	3.8	(3.7)	[3.8]	6.4	(6.3) [6.3]
Ireland						
Men	6.1	3.8	(3.3)	[3.6]	9.3	(8.9) [9.0]
Women	9.0	6.0	(5.3)	[5.8]	13.4	(12.7) [12.7]
Portugal						
Men	10.2	7.6	(7.4)	[7.3]	13.6	(13.1) [13.5]
Women	12.4	9.7	(9.6)	[9.4]	15.7	(15.3) [15.5]
Sweden						
Men	4.6	3.0	(2.8)	[2.8]	6.6	(6.3) [6.4]
Women	8.6	6.2	(6.0)	[6.1]	11.7	(11.2) [11.3]
Slovakia						
Men	4.2	2.5	(2.2)	[2.4]	6.7	(6.2) [6.1]
Women	5.7	4.1	(3.9)	[4.0]	7.8	(7.5) [7.8]
United Kingdom						
Men	8.6	6.7	(6.6)	[6.5]	11.0	(10.7) [10.7]
Women	9.0	7.0	(6.9)	[7.0]	11.5	(11.1) [11.2]

95 % empirical likelihood confidence intervals. The standard confidence intervals are given in round brackets. Rescaled bootstrap confidence intervals are given in squared brackets

Note that for Poland, the standard interval is shorter than the empirical likelihood and bootstrap intervals.

These confidence intervals are for illustrative purpose only, and are not part of any results officially released by Eurostat. The quality of these confidence intervals relies on the availability and quality of the design variables. These confidence intervals are likely to be conservative because the effect of calibration adjustment was not taken into account. This effect may be more pronounced for Scandinavian countries which relies heavily on calibration adjustment. The confidence intervals for Belgium may be too large because EU-SILC Belgium uses large sampling fractions.

4   Conclusions

The proposed empirical likelihood approach is a simple alternative for constructing confidence intervals. The proposed approach is based on several approximation: (a) the poverty threshold is assumed fixed, (b) it is assume that most of the variability is captured by the differences between the PSU totals. Note that the same approximation is used by Eurostat for variance estimation (Di Meglio et al. 2013). The drawback of the empirical likelihood approach is that it requires a specialised R library which is currently under development by the authors.

The proposed empirical likelihood approach does not rely on normality of the point estimator, variance estimates, linearisation, re-sampling and joint inclusion probabilities. The proposed approach is simpler than linearisation and naturally includes calibration. The empirical likelihood confidence intervals for the persistent-risk-of-poverty rate tends to be shifted upwards compared with the bootstrap and standard confidence intervals; this can be explained by the fact the empirical likelihood preserves the range of the data and takes into account of the distribution of the data.

The coverage of standard confidence intervals can be poor with skewed variables or cannot preserve the range of the data. For example, the persistent risk of poverty is always positive by definition. However we may have negative lower bounds for domains when we use standard confidence intervals based upon the normality (e.g. Austria, Denmark, Ireland and Malta in Table 4). Lower bounds are always positive for the empirical likelihood approach. Standard confidence intervals may also have poor coverage because they rely on normality which may not hold with skewed data. This is usually the case for EU-SILC poverty indicators.

References

Atkinson, A.B., Marlier, E.: Income and Living Conditions in Europe. Office for Official Publications, Luxembourg (2010). http://epp.eurostat.ec.europa.eu/cache/ITY_OFFPUB/KS-31-10-555/EN/KS-31-10-555-EN.PDF

Berger, Y.G.: Rate of convergence to normal distribution for the Horvitz-Thompson estimator. J. Stat. Plan. Inference **67**, 209–226 (1998)

Berger, Y.G., De La Riva Torres, O.: *An Empirical Likelihood Approach for Inference Under Complex Sampling Design.* Southampton Statistical Sciences Research Institute, Southampton (2012). http://eprints.soton.ac.uk/337688

Berger, Y.G., Skinner, C.J.: Variance estimation of a low-income proportion. J. R. Stat. Soc. Ser. C (Appl. Stat.) **52**, 457–468 (2003)

Chen, J., Sitter, R.R.: A pseudo empirical likelihood approach to the effective use of auxiliary information in complex surveys. Stat. Sin. **9**, 385–406 (1999)

Chen, J., Chen, S.R., Rao, J.N.K.: Empirical likelihood confidence intervals for the mean of a population containing many zero values. Can. J. Stat. **31**, 53–68 (2003)

Deville, J.C.: Variance estimation for complex statistics and estimators: Linearization and residual techniques. Surv. Methodol. **25**, 193–203 (1999)

Deville, J.C., Särndal, C.E.: Calibration estimators in survey sampling. J. Am. Stat. Assoc. **87**, 376–382 (1992)

Di Meglio, E., Osier, G., Goedemé, T., Berger, Y.G., Di Falco, E.: Standard Error Estimation in EU-SILC - First Results of the Net-SILC2 Project. Proceeding of the Conference on New Techniques and Technologies for Statistics, Brussels (2013). http://www.cros-portal.eu/sites/default/files/NTTS2013fullPaper_144.pdf

Eurostat (2012) European union statistics on income and living conditions (EU-SILC). http://epp.eurostat.ec.europa.eu/portal/page/portal/microdata/eu_silc. Accessed 7 Jan 2013 [Online]

Godambe, V.P.: An optimum property of regular maximum likelihood estimation. Ann. Math. Stat. **31**, 1208–1211 (1960)

Hájek, J.: Asymptotic theory of rejective sampling with varying probabilities from a finite population. Ann. Math. Stat. **35**, 1491–1523 (1964)

Hájek, J.: Comment on a paper by D. Basu. in Foundations of Statistical Inference. Holt, Rinehart and Winston, Toronto (1971)

Hansen, M., Hurwitz, W., Madow, W.: Sample Survey Methods and Theory, Vol. I. Wiley, New York (1953)

Hartley, H.O., Rao, J.N.K.: A new estimation theory for sample surveys, II. A Symposium on the Foundations of Survey Sampling Held at the University of North Carolina, Chapel Hill, North Carolina. Wiley-Interscience, New York (1969)

Horvitz, D.G., Thompson, D.J.: A generalization of sampling without replacement from a finite universe. J. Am. Stat. Assoc. **47**, 663–685 (1952)

Hudson, D.J.: Interval estimation from the likelihood function. J. R. Stat. Soc. **33**, 256–262 (1971)

Isaki, C.T., Fuller, W.A.: Survey design under the regression super-population model. J. Am. Stat. Assoc. **77**, 89–96 (1982)

Krewski, D., Rao, J.N.K.: Inference from stratified sample: properties of linearization jackknife, and balanced repeated replication methods. Ann. Stat. **9**, 1010–1019 (1981)

Osier, G.: Variance estimation for complex indicators of poverty and inequality using linearization techniques. Surv. Res. Method **3**, 167–195 (2009)

Owen, A.B.: Empirical Likelihood. Chapman & Hall, New York (2001)

Polyak, B.T.: Introduction to Optimization. Optimization Software, Inc., Publications Division, New York (1987)

Rao, J.N.K., Wu, C.F.J.: Resampling inference with complex survey data. J. Am. Stat. Assoc. **83**, 231–241 (1988)

Rao, J.N.K., Wu, W.: Empirical likelihood methods. In: Pfeffermann, D., Rao, C.R. (eds.) Handbook of Statistics: Sample Surveys: Inference and Analysis, vol. 29B, pp. 189–207. North-Holland, The Netherlands (2009)

Rao, J.N.K., Wu, C.F.J., Yue, K.: Some recent work on resampling methods for complex surveys. Surv. Methodol. **18**, 209–217 (1992)

Särndal, C.-E., Swensson, B., Wretman, J.: Model Assisted Survey Sampling. Springer, New York (1992)

Wilks, S.S.: Shortest average confidence intervals from large samples. Ann. Math. Stat. **9**, 166–175 (1938)

Wu, C.: Algorithms and R codes for the pseudo empirical likelihood method in survey sampling. Surv. Methodol. **31**, 239–243 (2005)

Wu, C., Rao, J.N.K.: Pseudo-empirical likelihood ratio confidence intervals for complex surveys. Can. J. Stat. **34**, 359–375 (2006)

Responsive Design for Economic Data in Mixed-Mode Panels

Annamaria Bianchi and Silvia Biffignandi

Abstract In order to estimate socio-economic variables in a survey, time and cost constraints must be taken into account. Economic statistics survey-based data nowadays could benefit in terms of quality and cost by implementing differential design strategies thanks to the availability of registers containing auxiliary variables, and the possibility of collecting and monitoring the data collection process by using IT data storage during the process. Our study investigates the potentials of the application of responsive design strategies in the collection of data for the estimation of economic indicators. Special attention is devoted to non-response bias reduction. Some thoughts on the consequences on total survey error (TSE) are also proposed. The empirical analysis is based on data from the recruitment of a mixed-mode panel for businesses, the PAADEL panel.

Keywords Balance • Data quality • Mixed-mode panel • Representativeness • Response rate • Responsive design

1 Introduction

The data quality of economic surveys is a relevant problem. One of the main issues is non-response. This can be a problem to survey quality as long as non-response causes systematic error (bias) in the survey estimates. In the fixed response model, non-response bias is determined by two factors: the response rate and the difference between respondents and non-respondents on survey statistics. A similar decomposition holds within the framework of the random response model (Bethlehem et al. 2011).

A. Bianchi • S. Biffignandi
DSAEMQ, University of Bergamo, via dei Caniana 2, 24127 Bergamo, Italy
e-mail: annamaria.bianchi@unibg.it

F. Mecatti et al. (eds.), *Contributions to Sampling Statistics*, Contributions to Statistics,
DOI 10.1007/978-3-319-05320-2_6,
© Springer International Publishing Switzerland 2014

Up to now generally, data collection has been guided by response rates only. This strategy could be effective in those situations where response rates are predictive of non-response bias. However, to the extent that the response rate is not a good indicator of non-response bias, using it as a guide to data collection may provide distorted results. Several papers find that greater efforts to increase or maintain response rates do not significantly improve final estimates, or, even increase non-response bias (Keeter et al. 2000; Curtin et al. 2000; Groves 2006; Heerwegh et al. 2007; Groves and Peytcheva 2008; Merkle and Edelman 2009).

This general consideration becomes more relevant if we consider the current situation where decreasing trends in response rates are observed. The corresponding increasing resources needed to achieve preset response rates may be hardly sustainable. Moreover, hunting respondents too much could affect the quality of the answers, giving no gains in terms of estimates: if the additional cases are similar to other cases that have already been collected, then this action may not reduce the risk of non-response bias.

It is therefore becoming clear that alternative measures capturing additional information about the composition of the responding sample should be used and pursued during data collection. It is worth noticing that this problem is crucial also in the framework of Official Statistics. Indeed, National Statistical Offices need to provide timely and high-quality estimates in the current situation characterized by low-budget constraints and decreasing response rates. Among alternative approaches in monitoring the sample during the data collection, different forms of responsive design have recently been proposed. The term was used by Groves and Heeringa (2006) to describe survey designs where indicators of cost and error are used to monitor data collection and to determine when design changes should be made on an ongoing basis during data collection. A related design uses prior knowledge in designing the survey. Wagner (2008) defines adaptive and dynamic designs as differential survey designs tailored to the characteristics of sample units. In this case subgroups to be specially targeted are identified before data collection begins.

The main idea underlying these methods is to intervene in the data collection process, in order to achieve an ultimate set of responding units that is "better balanced" or "more representative" than if no special effort is made. Two main issues can be identified. The first one refers to how units that should be targeted are identified. Generally, proper indicators are used. These indicators should be computable from the available data. Wagner (2012) classifies a number of possible indicators into three categories based on types of data used to estimate the indicators. The proposed classification is summarized in Sect. 2. The second issue is how the identified units should be targeted (e.g. incentives, different mode of communication, prioritization of cases, timing of the interviews, choice of the interviewers, etc). In this chapter, we focus on the first issue in the context of responsive design.

The recent existing literature presents much progress in the development of this methodology. However, further investigations are needed in order to apply it in practice, also to different contexts. The aim of this chapter is twofold: (a) to evaluate the potentials of responsive design in the estimation of a variable in the ongoing businesses probability-based PAADEL panel. This would be useful to revise the recruitment process in the future; (b) since the PAADEL panel is a mixed mode one, where one mode is Web, our study provides empirical findings on responsive design in this context. Further, it has to be stressed that up to now most of the empirical work on responsive/adaptive designs is carried out for large households/person surveys. It is important to evaluate the potentials of this technique also for business surveys and small surveys possibly with low response rates. These kinds of surveys may have their specific features which in turn may differentially influence the outcomes of differential designs. Thus our study represents also a contribution in the direction of the use of responsive design in business surveys.

To accomplish these goals, using the database of the PAADEL research, first the data collection process is analyzed. Next, a set of experimental responsive designs based on alternative interventions in the data collection is artificially reproduced. Results are analyzed in a comparative way to evaluate the impact of this approach on the final estimates. Special attention is devoted to non-response bias reduction. Some thoughts on the consequences on total survey error (TSE) are also proposed.

The chapter is organized as follows. Section 2 describes a typology of measures that have been used and implemented in responsive designs as well as a review of the literature on responsive design. Section 3 describes the PAADEL database, while Sect. 4 contains the results of our analyses. Section 5 discusses the findings and offers some conclusions.

2 Indicators and Responsive Design

As evidenced in the previous section, the first step towards the implementation of responsive/adaptive designs is the identification of units mostly contributing to non-response bias. These units should be specially targeted during data collection. The identification of such units is carried out through indicators, which represent proxy measures for non-response bias. Several indicators have been proposed in the literature. Wagner (2012) suggests an extension of the typology proposed by Groves et al. (2008). The proposed typology is based on the types of data used to estimate them. The indicators are the following:

1. *Involving the response indicator*, i.e., the response rate. The principal values of the response rate are its simplicity (a single-number summary for a survey) and its ease-of-use. These characteristics can explain its spread use. An implicit model is required to relate the non-response rate to the non-response bias. Generally, it is implicitly assumed that the higher the non-response rate, the higher the risk of non-response bias. As discussed in Sect. 1, both theory and empirical evidence show that these assumptions about the relationship between the response rate and non-response bias need not be true and that the response rate is not necessarily a good predictor of non-response bias.

2. *Involving the response indicator and frame data or paradata* (available both for respondents and non-respondents). Paradata are data about the data collection process (e.g., call records, observations made by interviewers, timing data, and mode of communication). Subgroup response rates are an example of Type 2 indicators. They involve the response indicator and a variable on the sampling frame or paradata that differentiates the sample into subgroups of interest. Other examples of Type 2 indicators are variance functions of non-response weights (Groves et al. 2008), variance functions of poststratification weights, variance functions of response rates on subgroups, balance indicators (Särndal 2011), goodness of fit statistics on propensity models (Särndal and Lundström 2008), and R-indicators (Schouten et al. 2009). Comparisons of survey data to published population totals fall into this category as well (Skalland 2011). Type 2 indicators are reported at the survey level. On one hand, this is a strength of these indicators, since it allows for a single number to characterize the entire survey. On the other hand, it is also a weakness, since the underlying assumption is that a set of auxiliary variables/paradata are able to capture non-response biases of all survey estimates. This may be an issue, especially in multipurpose surveys. It is known that this assumption is likely to be false, even though it is certainly more tenable than the implicit assumption underlying Type 1 indicators.

3. *Involving the response indicator, frame data or paradata, and the survey data* (observed for respondents only). Examples of Type 3 indicators are correlations between auxiliary data and survey variables (Kreuter et al. 2010), correlations between post-survey weights and the survey variables, the variations of means across the deciles of the survey weights, comparison of respondents means across deciles of estimated response propensities, comparison of late and early respondents, follow-up surveys of non-respondents, and the fraction of missing information (FMI) (Wagner 2010; Andridge and Little 2011). Since non-response bias occurs at the statistic level, a strength of these indicators is that they are reported at that level. However, this implies greater complexity for users of survey estimates. In this direction, Groves et al. (2008) propose compromises between indicators at the survey level and estimate-specific indicators (indicators of non-response errors for sets of estimates having similar correlations with different auxiliary variables).

A potential weakness of Type 2 and Type 3 indicators is that they are sensitive to model specification. Statements about the risk of non-response bias assume that these models are correctly specified. Misspecification of these models may lead to incorrect conclusions. Moreover, these indicators depend on the availability of good auxiliary variables, which should be related both to the response behavior and to key survey variables.

Using Type 1 indicators (response rates) to guide data collection leads survey organizations to struggle to attain preset values for the response rate. As empirical evidence has shown, this may lead to distorted practices. Since all cases contribute equally to the response rate, the most cost-effective action is to target cases that are most likely to respond. This is not the case for Type 2 and Type 3 indicators.

When these measures are used to guide data collection, survey organizations would not simply be interested in how many cases they interview, but also in which cases are interviewed. Under these measures, cases have different impacts on the guiding indicators and, therefore, can be given different priorities during data collection. An issue for Type 3 indicators is that they generally assume different values for different estimates in the same survey. This can make it difficult to use them to design a fieldwork strategy.

The exploitation of indicators for implementing responsive/adaptive designs is just beginning. Many open problems still remain.

Some attempts are based on subgroup response rates. The 2006–2010 US National Survey of Family Growth (NSFG) uses a management decision model to guide interventions in a responsive design framework (Lepkowski et al. 2013). Two responsive design features are implemented: interventions to improve response rates and two-phase sampling for non-response. Further, several experimental interventions in the fieldwork have been investigated. These interventions are reported and analyzed by Wagner et al. (2012). One of these experimental interventions uses the variation in subgroup response rates to restore balance in the composition of interviewed cases. The intervention consists in prioritizing cases from subgroups that are responding at lower rates. The responsive design has the beneficial effect of improving the composition of the final response set.

Mohl and Laflamme (2007) and Gambino et al. (2010) describe a responsive design strategy developed for computer-assisted telephone interviewing (CATI) surveys at Statistics Canada. The identification of cases is based both on response propensities and subgroup response rates.

Lundquist and Särndal (2013) report several experiments in the spirit of responsive design on the 2009 Living Condition Survey (LCS) at Statistics Sweden. The experiments are based on subgroup response rates. The results showed that appropriate interventions in the data collection can bring considerable improvements, while lowering costs, compared to the traditional LCS data collection.

To the extent that the characteristics used to balance the sample are predictive of the key statistics measured by the survey, equalizing response rates across subgroups defined by these characteristics is expected to reduce non-response bias. This approach corresponds to the "Missing At Random" (MAR) notion (Little and Rubin 2002). In the case of the NSFG, demographic variables used for balancing are predictive of key survey measures. However, this is not always the case. In a meta-analysis of 23 specialized studies of non-response, Peytcheva and Groves (2009) find no evidence that variation in response rates across groups defined by these types of demographic variables is related to biases in key survey variables. The topics of the surveys considered in the analysis range from health to employment, education, consumer satisfaction, voting behavior, parenthood, and lifestyle. We are not aware of similar studies for business surveys.

Three main drawbacks to using subgroup response rates in monitoring and intervening in the data collection process have been identified (Schouten et al. 2011). First, subgroup response rates do not account for subgroup size, i.e., small subgroups may appear equally important as large subgroups. Second, subgroup response rates are not available at the variable level. This implies that different variables cannot be evaluated and compared in their impact on response. Third, subgroup response rates are univariate and do not easily allow conditioning on other variables. In response to these drawbacks, Schouten et al. (2011) propose the use of partial R-indicators.

Shlomo et al. (2013) provide a simulation study where they demonstrate how a responsive design can be carried out using partial R-indicators. They show that even with a slight increase in the response rate, large gains in representativity can be obtained when targeting the data collection.

Luiten and Schouten (2013) use tailored adaptive design to obtain a more representative response set in an experimental setting for the Survey of Consumer Sentiment (SCS). Paradata from previous SCS and register information are used to identify subgroups with different contact and co-operation propensities by calculating partial R-indicators. The tailored design seeks to reduce the variability in response propensities among groups by assigning sample units to different modes in an initial approach, and by differentiating the timing and number of CATI contact attempts, and the interviewers assigned to specific sample units. The strategy is successful in significantly increasing representativeness, while maintaining the level of response and costs.

Recently, Schouten et al. (2013) proposed a general framework for optimizing response quality given cost constraints. The framework encompasses several one-dimensional quality functions.

In general, many approaches and variables can be adopted in the definition of the subgroups; how to define these subgroups and how alternative approaches impact on the data quality and costs is not trivial and further research on this issue is certainly needed.

Our study investigates the use of subgroup response rates to define a responsive design in a mixed-mode business survey.

3   The PAADEL Panel

The PAADEL-Producer panel is an Italian regional panel of businesses in the agro-food sector managed at the CASI Centre of the University of Bergamo.

The sample has been drawn from the Italian Business Statistical Register (ASIA, Archivio Statistico delle Imprese Attive). It is a stratified sample by size class and by activity sector. For large industries (over 250 employees)—due to the limited number—all the enterprises are considered. The sample size is 1,831 enterprises. The target population is the set of all enterprises operating in the food and beverage manufacturing sector (Ateco 10 and 11 in the Italian sectoral activity classification) in Lombardy.

The recruitment of the panel was conducted in 2012 and lasted approximately 3 months. The recruitment was conducted in two steps. The first recruitment step was based on phone mode (maximum number of contact attempts 5); the second

Table 1 Profiling rates by size class

	Rates			Percentage composition of recruitment		
	Food (Ateco 10)	Beverage (Ateco 11)	Total	Food (Ateco 10)	Beverage (Ateco 11)	Total
Class 1	23.4	37.5	24.0	69.0	92.3	70.0
Class 2	26.0	8.3	25.2	22.9	7.7	22.2
Class 3	29.2	0.0	26.9	4.9	0.0	4.7
Class 4	31.0	0.0	25.0	3.2	0.0	3.0
Total	24.4	23.5	24.4	100.0	100.0	100.0

step was based on the mixed mode approach (Web, phone, mail, fax). During the first phone contact a preliminary panel subscription was acquired and the request of the chosen mode for the questionnaire compilation in the second recruitment step (second step is related to questions for profiling the enterprise) was made. The recruitment rate of the first step is 66.5 %. The profiling rate of the second step is equal to 24.4 %: probably the fragmentation due to the vacation shifts had created interruption effects in the recruitment profiling process. The profiling rate of the second step by size class is displayed in Table 1.

During the recruitment process, the following paradata have been collected: number of contact attempts, selected survey mode, reasons for no contact or refusals. With regard to auxiliary variables, we have information on province, size (in number of employees), economic activity, and legal status. Moreover, in order to integrate the frame with information on local labour systems (LLS) and Industrial Districts (ID), we merged it with the Italian National Statistical Institute (Istat) data retrieved from the 2001 Census. LLSs are aggregated municipalities at an intermediate territorial level. This spatial partition was carried out by Istat in collaboration with the University of Parma on the basis of 2001 census data related to home-work commuting flows. We introduced a variable to identify businesses belonging to LLSs of large cities. IDs are identified by Istat on the basis of the LLSs of the 2001 Census. IDs are geographically concentrated areas of small and medium enterprises (SMEs). Enterprises in the district mainly belong to the same industrial sector, which constitutes the main industry. Out of 58 LLSs in Lombardy, 27 are identified as IDs.

4   Results

This section reports the results of experimental responsive design strategies applied to the PAADEL panel. The identification of units mostly contributing to non-response bias is based on the subgroup response rates. Even though this method has some drawbacks (Schouten et al. 2011), it proved to be quite effective in some cases (Wagner et al. 2012; Gambino et al. 2010). Further it is easier, quicker, and more intuitive to implement. We are interested to see how this method performs in our context. So, as a first experimentation, this seems to be the most appropriate one.

In the sequel, first the progression of the estimates of a few variables is studied as data collection proceeds. Next, the balance and representativity of the panel are investigated at different steps by means of subgroup response rates-based indicators. Finally, a set of responsive designs based on alternative interventions in the data collection is artificially reproduced. The results are analyzed both in terms of non-response bias and TSE perspectives. Some thoughts on the consequences of the mean squared error (MSE) are also addressed.

The analysis presented in this section refers to SME (with less than 250 employees).

4.1 Analysis of the Data Collection

The target variable used in the analysis is the legal status of the firm, in three categories: self-employment, partnership, and corporation. The choice of the target variable was determined by a number of reasons. First, it is a variable available in the frame as well as at the population level (from ASIA database). This allows the separate identification and investigation of the non-response error and the total error in the estimates. Second, the choice is based on the economic hypothesis that the legal form of a company may result in different business management behaviors, consequently leading to different answers for the variables detected in the survey. In this sense, a correct sampling strategy with respect to the legal form variable could have important implications on the collection of the substantive variables in the questionnaire.

For a given response set, design weights are calibrated on the population distribution of enterprises by province and by size × sectoral activity (Deville and Särndal 1992). These variables are similar to those used to produce the final calibration estimates in the panel. The population values used for calibration are obtained from the ASIA database. The calibration estimator produces the estimates denoted by $\widehat{\overline{y}}^C$. For comparison purposes, the expansion estimator $\widehat{\overline{y}}^E$ is computed as well. The expansion estimator is a calibration estimator calibrating only on the total number of units in the population. In order to isolate the error due to non-response, the estimates are compared to the Horwitz–Thompson estimator based on the entire sample $(\widehat{\overline{y}}^S)$. Moreover, to exclude possible measurement errors, the estimates are computed using values from the register variables and not the collected ones. The estimates are compared to the corresponding full sample values by means of the relative difference

$$\mathrm{RD}^C = \frac{\widehat{\overline{y}}^C - \widehat{\overline{y}}^S}{\widehat{\overline{y}}^S} \quad \mathrm{RD}^E = \frac{\widehat{\overline{y}}^E - \widehat{\overline{y}}^S}{\widehat{\overline{y}}^S}.$$

Table 2 Progression of the estimates and weighted response rates (P)

Attempt	P	Self-employed		Partnership		Corporations		Legal status	
		RD^E	RD^C	RD^E	RD^C	RD^E	RD^C	ARD^E	ARD^C
1	9.26	0.044	0.094	0.003	0.015	0.030	0.181	0.025	0.097
2	17.03	0.018	0.069	0.039	0.037	0.041	0.002	0.033	0.036
3	18.70	0.049	0.007	0.003	0.005	0.050	0.026	0.034	0.012
4	20.57	0.112	0.053	0.027	0.011	0.088	0.069	0.075	0.044
5	24.09	0.127	0.103	0.033	0.038	0.218	0.221	0.126	0.121

An aggregate measure (over all categories) is the average relative difference (ARD), defined as the mean of the relative differences over the three categories.

Table 2 contains the progression of the estimates for the target variables as a function of the number of contact attempts at which the sampled firm answered the recruitment questionnaire. For almost all variables and almost all attempts, the relative difference for the expansion estimator is larger than the one for the calibration estimator. The auxiliary variables used in the calibration are quite effective in reducing the estimation error. Exceptions are observed at the first and second contact attempts. Non-response error varies by the amount of data collected. The errors in the estimates are generally larger at the end of the data collection rather than at earlier stages. For example, for legal status, the ARD for the calibration estimator is 0.121 at the end of the data collection and 0.036, 0.012, and 0.044 at the second, third, and fourth attempts, respectively. In general, the results from the table suggest that the data collection for the PAADEL panel is not efficient enough and that possibly appropriately intervening in the data inflow could lead to better estimates.

In order to monitor the data collection we chose the Type 2 indicator subgroup response rates. The response rates may be divided into components using both auxiliary variables and paradata. To define the subgroups, we considered the auxiliary variable size and the paradata variable mode. Mode is known to produce different non-response mechanisms. Also different behavior is expected from firms of different sizes. The following subgroups have been considered:

- Size 0–9, online mode.
- Size 10–49, online mode.
- Size 50–249, online mode.
- Size 0–249, non-online mode.

The last category corresponds to a difficult-to-reach population. Enterprises that chose to answer the questionnaire non-online are expected to be rather different from the others and it is generally more difficult to contact and obtain cooperation from them. The choice of the variables and categories to be used has also been determined by the small number of observations. The inclusion of more variables/categories in the model would have implied the creation of cells with too small numbers of observations which in turn would make the results rather unstable.

Table 3 Response rates

Size	Mode	Attempt				
		1	2	3	4	5
0–9	Online	11.44	20.41	22.59	24.76	29.33
10–49	Online	8.26	15.73	17.69	18.48	24.39
50–249	Online	7.98	17.96	19.95	21.95	25.94
0–249	Non-online	6.70	12.72	13.56	15.41	16.45

Since the response rate in the first step of the recruitment has been rather high, while we had major problems during the second step, indicators are computed for the final response set with respect to the units contacted in the first step.

Table 3 contains the weighted (using design weights) response rates by a subgroup. The first group (small online enterprises) is the one which shows the highest response. The second and third groups have slightly lower response rates. As expected the hard-to-reach group is the one with the lowest response rates over the entire data collection. Probably more effort should be employed to obtain answers from this group.

Subgroup response rates can be aggregated in order to have a single indicator for the entire survey. Indicators of this type allow the detection of trends in data quality throughout the fieldwork. Groves et al. (2008) propose variance functions (e.g., the coefficient of variation CV) of the response rates on the subgroups. Lundquist and Särndal (2013) prove that by appropriately combining the subgroup response rates, balance indicators can be obtained. Balance indicators are defined as follows. Given an auxiliary vector $x_i = (x_{i1}, \ldots, x_{ik})^T$, let $\bar{x}_r$ and $\bar{x}_s$ be the weighted means for the response set and the full sample, respectively. Define $D = \bar{x}_r - \bar{x}_s$ and the Mahalanobis type distance $D^T \sum_s^{-1} D = (\bar{x}_r - \bar{x}_s)^T \sum_s^{-1} (\bar{x}_r - \bar{x}_s)$, where the weighting matrix is $\sum_s = \sum_s d_i x_i x_i^T / \sum_s d_i$, d_i being the survey weight for unit i. Two balance indicators on the unit interval have been introduced (Särndal 2011)

$$\text{BI}_1 = 1 - \sqrt{\frac{D^T \sum_s^{-1} D}{1/P - 1}} \quad \text{BI}_2 = 1 - 2P \sqrt{D^T \sum_s^{-1} D},$$

where P is the weighted response rate. Higher values for the indicators correspond to higher balance in the sample. Lundquist and Särndal (2013) prove that in case of crossed categorical variables, these indicators result indeed in a combination of response rates. In fact, they show that

$$D^T \sum_s^{-1} D = \sum_{j=1}^J W_{js} \left(\frac{P_j - P}{P}\right)^2,$$

where the sum is over the J categories of the categorical variable, P_j is the weighted response rate for group j, and W_{js} is the trait's share of the whole sample.

Table 4 Progression of indicators

Attempt	P	CV	BI$_1$	BI$_2$
1	9.26	23.417	0.925	0.956
2	17.03	19.600	0.907	0.930
3	18.70	20.734	0.895	0.918
4	20.57	20.207	0.894	0.914
5	24.09	22.720	0.864	0.884

Table 5 Strategy 1 determination

| Size | Mode | Attempt | | | | |
		1	2	3	4	5
0–9	Online	11.44	20.41	22.59	**24.76**	29.33
10–49	Online	8.26	15.73	17.69	18.48	**24.39**
50–249	Online	7.98	17.96	19.95	21.95	**25.94**
0–249	Non-online	6.70	12.72	13.56	15.41	16.45

Table 4 shows the progression of the CV of the subgroup response rates, and the balance indicators BI$_1$ and BI$_2$. Ideally, the balance indicators should increase over the data collection and the variance of the response rates should decrease. However, the table shows that the indicators go in the wrong direction: the balance is decreasing and the CV of the subgroup response rates is increasing.

4.2   Experiments

This section presents the results of different experimental responsive designs based on the subgroup response rates. First, three experiments in retrospect are carried out. The interventions aim at restoring balance in the composition of the interviewed cases. They consist in declaring data collection terminated for a certain subgroup as soon as its response rate reaches a certain predefined value.

Strategy 1 declares data collection terminated in a subgroup as soon as the response rate 24 % is reached. Table 5 shows how the experimental strategy is determined. The groups for which data collection is terminated at a certain contact attempt. So, according to Strategy 1, data collection is terminated for group 1 (size 0–9, online mode) at the fourth contact attempt. Data collection continues till the end for the other subgroups. The preset response rate 24 % is reached by the end of the data collection for the second (size 10–49, online mode) and the third (size 50–249, online mode) groups. The response rate for the hard-to-reach group (size 0–249, non-online mode) at the end of the data collection is only 16.45 %.

Table 6 shows the progression of the indicators over the data collection for Strategy 1. With respect to the values in Table 4, we expect to observe changes in the indicators only in the last step of the data collection, where the intervention

Table 6 Progression of indicators for Strategy 1

Attempt	P	CV	BI$_1$	BI$_2$
1	9.26	23.417	0.925	0.956
2	17.03	19.600	0.907	0.930
3	18.70	20.734	0.895	0.918
4	20.57	20.207	0.894	0.914
5	24.09	18.973	0.905	0.921

Table 7 Strategy 2 determination

| Size | Mode | Attempt | | | | |
		1	2	3	4	5
0–9	Online	11.44	**20.41**	22.59	24.76	29.33
10–49	Online	8.26	15.73	17.69	18.48	**24.39**
50–249	Online	7.98	17.96	19.95	**21.95**	25.94
0–249	Non-online	6.70	12.72	13.56	15.41	16.45

Table 8 Progression of indicators for Strategy 2

Attempt	P	CV	BI$_1$	BI$_2$
1	9.26	23.417	0.925	0.956
2	17.03	19.600	0.907	0.930
3	18.70	17.498	0.919	0.938
4	20.57	14.787	0.940	0.954
5	24.09	16.018	0.933	0.947

Table 9 Strategy 3 determination

| Size | Mode | Attempt | | | | |
		1	2	3	4	5
0–9	Online	11.44	**20.41**	22.59	24.76	29.33
10–49	Online	8.26	15.73	17.69	**18.48**	24.39
50–249	Online	7.98	17.96	**19.95**	21.95	25.94
0–249	Non-online	6.70	12.72	13.56	15.41	16.45

took place. As the indicators demonstrate, the balance of the response set has increased in the last step, reflecting the greater similarity among the response rates.

Strategy 2 considers data collection terminated for a subgroup when its response rate has reached 20 %. The strategy together with its intervention points is illustrated in Table 7.

Table 8 shows the progression of the indicators under Strategy 2. An improvement is observed with respect to Strategy 1. Changes in the indicators start from attempt number 3, where the first intervention is made.

The stopping rule for Strategy 3 is defined by declaring data collection terminated as soon as its response rate has reached 18 %. Strategy 3 is described in Table 9.

Table 10 shows the progression of the indicators under Strategy 3. More pronounced improvements are observed. For example, at the end of the data

Table 10 Progression of indicators under Strategy 3

Attempt	P	CV	BI_1	BI_2
1	9.26	23.417	0.925	0.956
2	17.03	19.600	0.907	0.930
3	18.70	17.498	0.919	0.938
4	20.57	12.166	0.942	0.955
5	24.09	9.486	0.954	0.964

Table 11 Estimates for the three experimental strategies

	Self-employed		Partnership		Corporation		Legal status	
Data	RD^E	RD^C	RD^E	RD^C	RD^E	RD^C	ARD^E	ARD^C
Full data set	0.127	0.103	0.033	0.038	0.218	0.221	0.126	0.121
Strategy 1	0.119	0.078	0.043	0.035	0.218	0.176	0.127	0.096
Strategy 2	0.026	0.022	0.100	0.086	0.202	0.142	0.109	0.083
Strategy 3	0.024	0.025	0.061	0.062	0.055	0.082	0.047	0.056

Table 12 Indicators for the three experimental strategies

Data	P	BI_1	BI_2
Full data set	0.241	0.816	0.843
Strategy 1	0.219	0.854	0.879
Strategy 2	0.196	0.863	0.891
Strategy 3	0.188	0.867	0.896

collection the balance indicator BI_1 reaches the value 0.954. This result could be expected, since lowering the target value for response rate allows the subgroup response rates to be more similar.

The fact that the indicators considered up to now improve when applying the strategies above was somehow expected since the indicators are summary measures of the subgroup response rates which become more similar under the three strategies.

It is interesting now to investigate the effects of the responsive design strategies on the final estimates as well as on the indicators based on a wider set of variables.

Table 11 compares the final estimates of the target variables for the full data set without interventions (already shown in Table 2) with those obtained under the three strategies. On the whole, the results improve. The ARD for the variable Legal status is reduced from one strategy to the next. The lowest error is observed for Strategy 3. A similar pattern applies to the variables Self-employment and Corporation. For the variable Partnership the better strategy seems to be the first one. The non-response bias reduction however is not uniform across the variables.

Table 12 contains the values of the balance indicators based on a wider set of variables (Size × Mode + Sectoral activity + Province + LLS of large city + Industrial district), together with the new response rates. The balance indicators assume lower values with respect to those computed before as they take into account more variables. They improve from one strategy to the next.

Table 13 Impact of the proposed strategies on TSE

Data	Self-employed		Partnership		Corporations		Legal status	
	RD^E	RD^C	RD^E	RD^C	RD^E	RD^C	ARD^E	ARD^C
Full data set	0.359	0.323	0.181	0.145	0.277	0.308	0.272	0.259
Strategy 1	0.360	0.309	0.179	0.156	0.286	0.265	0.275	0.243
Strategy 2	0.306	0.247	0.113	0.091	0.309	0.263	0.243	0.200
Strategy 3	0.271	0.242	0.161	0.127	0.131	0.177	0.187	0.182

Table 14 Standard error (SE) and mean squared error (MSE) for the calibration estimator

Data	Self-employed		Partnership		Corporations	
	SE	MSE	SE	MSE	SE	MSE
Full data set	3.19	182.21	3.48	42.25	2.93	44.82
Strategy 1	3.38	168.89	3.65	48.15	2.99	35.63
Strategy 2	3.60	113.37	3.79	26.30	3.13	36.08
Strategy 3	3.64	110.35	3.86	37.99	3.21	22.27

In order to evaluate the impact of the proposed strategies on the TSE, we compare the estimates to the corresponding population values. So, instead of $\widehat{\overline{y}}^S$, we use $\overline{y}^{POP}$ in the computation of the relative differences and ARDs. Moreover, the estimates are computed using values from the collected data (not frame data, as before). In this way, the relative differences include not only the effects of non-response, but also the effects of sampling error, noncoverage, and measurement error. The results are shown in Table 13.

As expected the error is greater than before. However, the implementation of the responsive design strategies has a positive impact on TSE as well.

Another interesting aspect related to responsive design regards the consequences of the application of such designs on the variance of the estimators. On one hand a smaller response rate means a smaller sample size and hence a higher variance. On the other hand, removing imbalance between respondents and non-respondents during data collection may be expected to reduce the variation in non-response adjustment weights and thus the variability of the estimates. Up to now, international research on responsive and adaptive survey designs has mainly focused on bias rather than variance issues. Beaumont and Haziza (2011) elaborate adaptive survey designs aiming at minimizing the variance of an estimator rather than its bias.

In order to throw some light on this issue, we computed standard errors (SE) and MSEs for the calibration estimator under different strategies. The results are shown in Table 14.

From the table, we see that in this case standard errors of the estimates slightly increase from one strategy to the next. However, in general MSEs decrease since in this case the main component of MSE is the bias of the estimates rather than the variance.

5 Conclusions

In order to estimate socio-economic variables in a survey, time and cost constraints must be taken into account. Nowadays, these issues are getting more and more crucial, due to increasing budget constraints and decreasing response rates. Recently, responsive and adaptive designs have been proposed to improve quality of the estimates and reduce costs. These issues are of great relevance also for Official Statistics. Indeed, many National Statistical Institutes such as Statistics Netherlands, Statistics Sweden, Statistics Canada and Istat are currently researching or planning to research in this area. The main advantages of these designs are the following:

(a) They may be viewed as extensions of sampling designs. Whereas standard sampling designs assume a uniform data collection strategy, responsive designs employ multiple strategies.
(b) They extend quality objective function. Sampling designs typically focus on precision, but responsive survey designs also attempt to account for non-sampling errors through indirect indicators.
(c) They allow for an explicit trade-off between quality and costs and other constraints. As such they need not restrict attention to non-response error, even though up to now most studies limited themselves to non-response.

In this chapter we apply experimental responsive design strategies to the recruitment of a mixed-mode panel of enterprises. The method proves to be promising in the reduction of non-response bias. When evaluated in the TSE framework, the results are not so clear-cut or uniform across variables. A reduction of the total error is observed in general. However, the pattern is rather different across different variables.

These conclusions should be taken with caution. Still more research is needed to deepen these results in several directions. For example, the problem of how to choose the value for the preset response rate would deserve more attention. Moreover, investigations on the consequences of applying responsive designs on the variability of the estimates are needed. Another interesting aspect is whether responsive and adaptive survey designs are ignorable. This is relevant whenever statistical models are used. If the sampling design is not ignorable, then analytical inference for superpopulation parameters is more difficult than in the ignorable case.

Acknowledgments The paper is supported by the ex-60 % University of Bergamo, Biffignandi grant and PAADEL project (Lombardy Region—University of Bergamo-CASI Center joint project). The authors thank Statistica e Osservatori, Éupolis Lombardia, for providing detailed population statistics. The authors would like to acknowledge networking support by Cost Action IS1004. The authors are thankful to the anonymous referee for valuable comments and remarks.

References

Andridge, R.R., Little, R.J.A.: Proxy pattern-mixture analysis for survey nonresponse. J. Official Stat. **27**, 153–180 (2011)

Beaumont, J.-F., Haziza, D.: A theoretical framework for adaptive collection designs. In: Paper Presented at 5th International Total Survey Error Workshop, June 21–23, Quebec (2011)

Bethlehem, J., Cobben, F., Schouten, B.: Handbook of Nonresponse in Household Surveys. Wiley, Hoboken (2011)

Curtin, R., Presser, S., Singer, E.: The effects of response rate changes on the index of consumer sentiment. Pub. Opin. Q **64**, 413–428 (2000)

Deville, J.C., Sarndal, C.E.: Calibration estimators in survey sampling. J. Am. Stat. Assoc. **87**, 376–382 (1992)

Gambino, J., Laflamme, F., Wrighte, D.: Responsive design at statistics Canada: development, implementation and early results. In: 21st International Workshop Household Survey Non-response, Nürnberg, August 30th–September 1st (2010)

Groves, R.M.: Nonresponse rates and nonresponse bias in household surveys. Pub. Opin. Q. **70**, 646–675 (2006)

Groves, R.M., Heeringa, S.G.: Responsive design for household surveys: tools for actively controlling survey errors and costs. J. R. Stat. Soc. A **169**, 439–457 (2006)

Groves, R.M., Peytcheva, E.: The impact of nonresponse rates on nonresponse bias: a meta-analysis. Publ. Opin. Q. **72**, 167–189 (2008)

Groves, R.M., Brick, J.M., Couper, M., Kalsbeek, W., Harris-Kojetin, B., Kreuter, F., Pennell, B.E., Schouten, B., Smith, T., Tourangeau, R., Bowers, A., Jans, M., Kennedy, C., Levenstein, R., Olson, K., Peytcheva, E., Ziniel, S., Wagner, J.: Issues facing the field: alternative practical measures of representativeness of survey respondent pools. Surv. Pract. **1**(3), 1–6 (2008)

Heerwegh, D., Abts, K., Loosveldt, G.: Minimizing survey refusal and noncontact rates: do our efforts pay off? Surv. Res. Methods **1**, 3–10 (2007)

Keeter, S., Miller, C., Kohut, A., Groves, R.M., Presser, S.: Consequences of reducing nonresponse in a national telephone survey. Pub. Opin. Q. **64**, 125–148 (2000)

Kreuter, F., Olson, K., Wagner, J., Yan, T., Ezzati-Rice, T.M., Casas-Cordero, C., Lemay, M., Peytchev, A., Groves, R.M., Raghunathan, T.E.: Using proxy measures and other correlates of survey outcomes to adjust for nonresponse: examples from multiple surveys. J. R. Stat. Soc. A **173**, 389–407 (2010)

Lepkowski, J.M., Mosher, W.D., Groves, R.M.: Responsive design, weighting, and variance estimation in the 2006–2010 National Survey of Family Growth. National Center for Health Statistics. Vital Health Stat. **2**(158) (2013)

Little, R.J.A., Rubin, D.B.: Statistical Analysis with Missing Data. Wiley, Hoboken (2002)

Luiten, A., Schouten, B.: Tailored fieldwork design to increase representative household survey response: an experiment in the survey of consumer satisfaction. J. R. Stat. Soc. A **176**, 169–189 (2013)

Lundquist, P., Särndal, C.E.: Aspects of responsive design with applications to the Swedish Living Conditions Survey, J. Official Stat. **29**, 557–582 (2013)

Merkle, D.M., Edelman, M.: An experiment on improving response rates and its unintended impact on survey error. Surv. Pract. **2**(3), 1–5 (2009)

Mohl, C., Laflamme, F.: Research and responsive design options for survey data collection at Statistics Canada. In: Proceedings of the American Statistical Association, pp. 2962–2968 (2007)

Peytcheva, E., Groves, R.M.: Using variation in response rates of demographic subgroups as evidence of nonresponse bias in survey estimates. J. Official Stat. **25**, 193–201 (2009)

Särndal, C.E.: The 2010 Morris Hansen lecture: dealing with survey nonresponse in data collection, in estimation. J. Official Stat. **27**, 1–21 (2011)

Särndal, C.E., Lundström, S.: Assessing auxiliary vectors for control of nonresponse bias in the calibration estimator. J. Official Stat. **24**, 167–191 (2008)

Schouten, B., Cobben, F., Bethlehem, J.: Indicators for the representativeness of survey response. Surv. Methodol. **35**, 101–113 (2009)

Schouten, B., Shlomo, N., Skinner, C.: Indicators for monitoring and improving representativeness of response. J. Official Stat. **27**, 231–253 (2011)

Schouten, B., Calinescu, M., Luiten, A.: Optimizing quality of response through adaptive survey designs. Surv. Methodol. **39**, 29–58 (2013)

Shlomo, N., Schouten, B., de Heij, V.: Designing adaptive survey designs with R-indicators. In: Paper Presented at the New Techniques and Technologies for Statistics (NTTS) Conference in Brussels, Brussels (2013)

Skalland, B.: An alternative to the response rate for measuring a survey's realization of the target population. Pub. Opin. Q. **75**, 89–98 (2011)

Wagner, J.: Adaptive survey design to reduce non-response bias. PhD Thesis, University of Michigan (2008)

Wagner, J.: The fraction of missing information as a tool for monitoring the quality of survey data. Pub. Opin. Q. **74**, 223–243 (2010)

Wagner, J.: A comparison of alternative indicators for the risk of nonresponse bias. Pub. Opin. Q. **76**, 555–575 (2012)

Wagner, J., West, B.T., Kirgis, N., Lepkowski, J.M., Axinn, W.G., Ndiaye, S.K.: Use of paradata in a responsive design framework to manage a field data collection. J. Official Stat. **28**, 477–499 (2012)

Comparing the Efficiency of Sample Plans for Symmetric and Non-symmetric Distributions in Auditing

Paola M. Chiodini and Mariangela Zenga

Abstract The aim of the sampling in auditing fields is to produce an upper confidence limits for mean or totals of unobserved mistakes on the observed values accounts (called *book values*) that are highly positively skewed. In the main auditing scenarios, it is possible to meet populations with a relatively large number of small accounts combined with high rate of mistakes, that means diffused low errors, or populations with a relatively large number of accounts combined with a small rate of mistakes, that means rare high errors. It seems interesting to evaluate whether the sampling method in relation to the different distribution shape of the population is able to distinguish the pure random error from the *systematic* error. In this work we will compare, by simulation, the mostly used audit sampling methods for different distribution shapes for book values with different hypothesis on the presence of the errors on the book values population.

Keywords Audit sampling • Monetary unit sampling • Stringer bound • Unequal inclusion probabilities

1 Introduction

Because of cost issues, during the practice of auditing, auditors only verify a sample of accounts coming from different sources and different means. There are two main types of test that can be based on statistical sampling in auditing practice. The first refers to the verification that internal procedures are followed; this is the *internal auditing*. The second refers to verification that reported monetary balances of large numbers of individual items are not *materially misstated*. This paper will discuss

P.M. Chiodini • M. Zenga (✉)
Department of Statistics and Quantitative Methods, University of Milano-Bicocca, Milano, Italy
e-mail: mariangela.zenga@unimib.it

F. Mecatti et al. (eds.), *Contributions to Sampling Statistics*, Contributions to Statistics,   103
DOI 10.1007/978-3-319-05320-2_7,
© Springer International Publishing Switzerland 2014

on the second type of test. The problem to be faced by the auditor is to decide, on sample basis, if an error can be considered *material* that is if its magnitude "... is such that it is probable that the judgement of a reasonable person relying upon the report would have been changed or influenced by the inclusion or correction or the item..." (FASB 1980). Different statistical problems should be faced when analyzing data coming from different non-standard mixture of distributions. An item in an audit sample produces two types of information: the *book* (recorded) amount and the *audited* (correct) amount. By the first two information it is possible to obtain a third one: the *error amount* that is difference between the book and the audited amount. The percentage of items in error in the sample can be very variable. It is common to observe only few items with errors, in this case the classical interval estimation of the total error amount based on the assumption of normality of the sampling distribution is not reliable. Again even when the sample does not contain any items in error, the estimated standard deviation of the estimator of the total error amount becomes zero. Generally speaking when the sample does not contain any item in error the difference between the estimated total audited amount and the book balance must be interpreted as pure sampling error. The auditor will evaluate the book amount free of error. As Stringer (1963) says on statistical estimation of the total error "... assuming a population with no error in it, each of the possible distinct samples of a given size that could be selected from it would result in a different estimate and precision limit under this approach; however, form the view point of the auditor, all samples which include no errors should result in identical evaluation...". Stringer proposed a statistical method for auditing procedures that did not depend on the normal sampling distribution and that could still provide a reasonable inference for the population error amount when all items in the sample are error free. This sampling method is the implementation of the monetary (dollar) unit sampling (MUS or DUS). Common statistical methods to select an audit sample are sampling without replacement (SRS) or the sampling technique known as "probability proportional to size" (PPS) without replacement. A particular case of the PPS is the MUS (Arens and Loebbecke 1981), which is popular because it directs the efforts towards high-valued items which contain the greatest potential of large overstatement.

Frequently auditors apply this method without any theoretical support for the efficiency of the result in relation to the audit situation. For the auditors the only interest is to verify that the total error falls within a pre-assigned value. If the observed total error is higher than the pre-assigned value a census is made on the contrary the auditors will conclude that no misstatement has been made. This means that there is no real interest to estimate the total error amount. This lack of interest is very practical: an accounting population may contain a relatively large number of small account combined with high rate of mistakes that is diffused small errors, or relatively large number of accounts combined with a small rate of mistakes that is rare big errors. Again, the auditors' choices in terms of sampling strategies are frequently based on their personal experience, that means they ignore the characteristic of the account population that generally speaking is positively skewed. In this paper we want to compare the performance of two mostly used sampling methods in auditing that is simple random SRS and MUS[1] in relation to different shapes of the account population combined with different error rates and different distributions of the errors. A real account population is used to simulate different scenarios. This paper is organized as follows. Section 2 describes the statistical sampling method mostly used in auditing and Sect. 3 presents the monetary unit sampling (MUS). The results of the simulations are presented in Sect. 4 and Sect. 5 offers the conclusions.

2   Statistical Sampling Method in Auditing

In audit sampling the following probability sampling methods (WOR) are used: simple random, stratified random and PPS. It is known that the PPS was developed by Hansen and Hurwitz within the survey sampling theory (Hansen and Hurwitz 1943) for the selection of cluster of unequal size. In audit sampling, this method is referred to as MUS, also known as dollar unit sampling (DUS). An account balance can be regarded as a cluster of monetary units being either correct or incorrect.

An accounting population consist of N line items with book (or recorded) values, $y_1, y_2, \ldots, y_N$ and total book amount T_y defined by:

$$T_y = \sum_{i=1}^{N} y_i. \tag{1}$$

The audited (true) amount of the N line items in the population is denoted by $x_1, x_2, \ldots, x_N$ and the total audited amount is:

$$T_x = \sum_{i=1}^{N} x_i. \tag{2}$$

The error in item i is $z_i = y_i - x_i$, $1 \leq i \leq N$. When $z_i > 0$, the i-th item is said to be overstated, and when $z_i < 0$, it is said to be understated. When $z_i = 0$, the account is said to be error free. The total error amount is defined as:

$$T_z = \sum_{i=1}^{N} z_i = \sum_{i=1}^{N} t_i \times y_i. \tag{3}$$

For $y_i \neq 0$, $t_i = \frac{z_i}{y_i} = \frac{y_i - x_i}{y_i}$ is called the fractional error or *taint* that is the ratio of error per dollar.

---

[1]The selection procedure for both sampling methods was the systematic procedure (Modow 1949).

The values $(x_1, x_2, \ldots, x_N)$ are unknown before sampling, whereas $(y_1, y_2, \ldots, y_N)$ are known. It is assumed that the amount of any overstatement does not exceed the stated recorded value. The purpose is to estimate, on sample basis, the total error amount T_z as the auditors are interested in obtaining an upper bound on T_z at a specified confidence level. If the upper bound exceeds the tolerable error amount, the auditor can decide that there is a material error in the book values, or on the contrary he can decide that the book values are simply affected by pure random error. Since a great proportion of the items in the population usually are error free while the non-zero errors are highly skewed to the right (Johnson et al. 1981; Neter et al. 1985), it is important to know the characteristics of the errors in order to determine the most appropriate and efficient sampling method.

The estimators commonly used in survey sampling such as Horvitz–Thompson (Horvitz and Thompson 1952), difference, ratio or regression estimator were first used in the estimation of the total error amount. In the auditing context the Horvitz–Thompson estimator with simple random SRS is referred to as the mean-per-unit. By performing a simulation study, Kaplan (1973) observed the behaviour of the mean per-unit, difference, ratio, and regression estimators at different sample sizes at 95 % confidence level. The main result is that the sampling distribution of the estimators often deviates from the Normal or Student's t distribution. Therefore the estimates of the mean were highly correlated with the estimates of the standard error. He suggested that a high correlation between the mean of mean-per-unit, difference, ratio, and regression estimators and standard error estimates would inflate the probability of a type II (type I) error.

Furthermore, several authors (Stringer 1963; Kaplan 1973; Neter and Loebbecke 1975, 1977) have noticed that since auditing populations are highly positively skewed, the small-value items have a high frequency compared to the large-value line items. Moreover the lack of normality of the error amounts distribution inhibits the use of the classical inference theory based on normality (Ramage et al. 1979; Johnson et al. 1981; Neter et al. 1985; Ham et al. 1985).

Several statistical methods have been developed in order to overcome the problems related to populations with low error rates and the limitations of the classical methods related to the normality assumption. These methods are based on sampling with probability proportional to recorded monetary value. These estimation methods are commonly referred to as combined attribute and variable (CAV) estimation (Goodfellow et al. 1974).

3   Monetary Unit Sampling

As the auditor's goal is to estimate the total amount of error through sample information, the MUS is the statistical sampling method more frequently used in practice. Despite its widespread use, the relative performance of the MUS compared to the traditional normal distribution variables is often not clear (Smieliauskas 1986). Many researchers have developed a number of methods for evaluating the

MUS samples (Tsui et al. 1985; Smieliauskas 1986; Grimlund and Felix 1987) by evaluating sample size (Kaplan 1975), sampling risks (Smieliauskas 1986), sample size implications of controlling for the same level of sampling risks (Smieliauskas 1986) and bounds (Tsui et al. 1985; Dworin and Grimlund 1984). It is important for the auditors to obtain reliable bounds on the total error in the population in order to take decisions at different confidence levels and probabilistic statements. Several methods are possible in order to compute bounds for the MUS sampling (Felix et al. 1981; Swinamer et al. 2004). The Stringer bound, introduced by Stringer (1963), is extensively used by auditors (Bickel 1992; AICPA 2008). The Stringer bound, which is particularly attractive to auditors, provides a non-zero upper bound even when no errors are observed in the sample. Simulations show that the Stringer bound reliably exceeds the true audit error (Swinamer et al. 2004). This reliability is favorable to auditors who are concerned with strictly overstatements (or strictly understatements) in financial statements. The Stringer bound is also simple to compute and the required statistical tables are easily available.

3.1   The Estimators of the Total Error Amount

Inferential methods based on normal distribution of the estimators do not seem to work. Indeed, two kind of problems are encountered when this sampling and estimation approach are applied to auditing:

- The first problem comes when there is the *Zero-Error Sample*: often accounting populations have very low error rates and consequently the selected sample may yield zero errors and hence fail to give any information on the real population total error amount that will be estimated at zero. In this case inferential method based on normality will fail to give confidence limit.
- The second problem is related to the *Unreliable Confidence Bounds*: when the average line item error amount is used as an estimate for the total error amount and the central limit theorem is applied to obtain the confidence limits, unreliable confidence intervals are constructed when the populations have low error rates and when the line items are highly skewed.

The interest of the auditors usually focuses on obtaining an upper bound on T_z, at a specified confidence level as accurate as possible. If this upper bound exceeds the tolerable error amount there is statistical evidence of a possible material error. When the upper confidence bound computed does not exceed the tolerable error amount, it is possible to decide that the recorded values are a fair reflection of the accounts.

In literature the upper confidence limit for T_z is used to evaluate the performance of the sampling method. Since estimators of T_z based on the large-sample normal distribution theory have been found to have a lower coverage than the stated confidence level (Kaplan 1973; Neter and Loebbecke 1975; Beck 1980), auditors frequently use heuristic non-classical bound estimates to determine the accuracy of financial statements (Horgan 1996).

The Stringer bound is one of the first CAV methods that were developed and still is the most widely used one. It was introduced by Stringer (1963) and it is a heuristic evaluation method. Let us assume that a sample of size n items selected without replacement is taken and let t_i be the taint value in line item i. If a selected monetary unit falls in the ith item, than the tainting t_i of the item is recorded. Namely, every monetary unit observation is the tainting of the item that the unit falls in. The book values are considered as realizations of an auxiliary variable, which is generally skewed. Let $T_1, \ldots, T_n$ be the independent random variables which represent the taintings. The distribution of these taintings is some unknown mixture of distributions within the interval $[0; 1]$, so that $Pr(0 \le T_i \le 1) = 1$. We denote $\mu = E(T_i)$ and let $0 \le t_{1:n} \le t_{2:n} \le \cdots \le t_{n:n} \le 1$ be the ordered statistics of $(T_1, T_2, \ldots, T_n)$. For $\alpha \in (0; 1)$ and $i = 0, 1, \ldots, n-1$ let $p = p_n(i; 1 - \alpha)$ be the unique solution of:

$$\sum_{k=0}^{i} \binom{n}{k} p^k (1 - p)^{n-k} = \alpha \tag{4}$$

with $p_n(n; 1 - \alpha) = 1$.

The Stringer method for obtaining an upper bound for the total overstatement error can be obtained by combining the upper limits for the sample error rates with the estimated taints:

$$\hat{T}_z = T_y p_n(0, 1 - \alpha) + T_y \sum_{i=1}^{n} [p_n(i, 1 - \alpha) - p_n(i - 1, 1 - \alpha)] t_{n-i+1:n} \tag{5}$$

then $p_n(i, 1 - \alpha)$ is the upper $(1 - \alpha)$ confidence limit for the binomial parameter when i errors are observed in a sample of size n. Equivalently for a given α, n and number of errors i, it is possible to find the value $p_n(i, 1 - \alpha)$ that satisfies:

$$\sum_{j=0}^{i} \binom{n}{j} [p_n(i, 1 - \alpha)]^j [1 - p_n(i, 1 - \alpha)]^{n-j} = \alpha. \tag{6}$$

The Stringer bound is sometimes calculated using the Poisson approximation for obtaining the upper confidence limits $p_n(i, 1 - \alpha)$.

This bound has been largely investigated and much empirical evidence supports the fact that its coverage is at least the nominal level. Moreover it has been found to be overly conservative (Leitch et al. 1982; Reneau 1978; Anderson and Teitlebaum 1973; Wurst et al. 1989) which means that the value of this bound is much larger than the real error amount. This conservatism can lead to the rejection of acceptable accounting populations (Leitch et al. 1982). Bickel (1992) has studied the asymptotic behaviour of the bound and has shown that for a large sample the confidence level achieved by the Stringer bound is very often higher than the nominal level. Pap and van Zuijlen (1996) have extended Bickel's work by demonstrating the asymptotic conservatism of the Stringer bound. They have shown that the bound is asymptotically conservative for confidence level $1 - \alpha$, when $\alpha \in (0, 1/2]$ and asymptotically not conservative when $\alpha \in (1/2, 1)$.

Many different less conservative methods than the Stringer bound have been proposed, the most famous are the multinomial bound, the cell bound, the moment bound and the bound based on Hoeffding's inequality. The multinomial bound was introduced by Fienberg et al. (1977). Their method is based on the multinomial distribution for evaluating MUS. The main difficulty is the complexity of finding the joint confidence region and maximizing over it, and this has made the method unpopular.

The cell bound was proposed by Leslie et al. (1979). This bound was developed for PPS with cell selection, but can also be used with unrestricted or systematic selection. Unfortunately for the auditing populations containing low error amount, the bound may be substantially greater than the actual error amount (Plante et al. 1985).

The moment bound was introduced by Dworin and Grimlund (1984, 1986). They constructed an upper bound for the total error amount by approximating the sampling distribution of the mean prorated error with a three-parameter gamma distribution. Using the sample moments, together with a heuristic approximations, they obtained estimates of the parameters of gamma distribution. Simulation studies suggest that the nominal coverage is close to the required one. Their results indicate that the moment bound is less conservative than the Stringer bound, and would correctly accept more accounts than the multinomial bound when the materiality limit is less than 4 % of the total recorded value. However, for materiality limits in excess of 5 %, the bound correctly accepted more accounts than the moment bound. The moment bound shows sporadic coverage failures (Grimlund and Felix 1987).

Hoeffding's inequality can be used to obtain confidence bounds for the mean prorated error. Fishman (1991) has described a confidence interval-estimation procedure based on this idea but the bound is more conservative than the Stringer bound.

Howard (1994) has proposed a combined bound based on a combination of a bootstrap approximation of the bound generated by Hoeffding's inequality and a modified nonparametric bootstrap-t method. The combined bound is not uniformly better than the Stringer bound. However, it seems to offer a moderate advantage over the Stringer bound for accounts with low error rates, low proportion of 100 % overstatement and high proportion of understatement.

Kvanli et al. (1998) have suggested that if the non-zero values in given auditing data can be assumed to come from some appropriate parametric model, then the likelihood ratio test can be used to construct a two-sided approximate confidence interval for the mean error amount. Their results indicate that the coverage attained by this method is very close to nominal when the auditing data follows the assumed parametric model. The limit of this procedure is that it depends on the choice of the underlying model.

Table 1 Characteristics of a
real accounting population

Characteristic	Value
Total line-items	188
Minimum	17.97 €
1st quartile	344.43 €
Mean	5,529.77 €
Median	1,000.33 €
3rd quartile	3,100.80€
Maximum	134,157.42 €
Standard deviation	16,172.41 €
Total recorded value	1,034,066.63 €
Skewness	5.33
Kurtosis	31.99

Fig. 1 The histogram for the
real accounting population

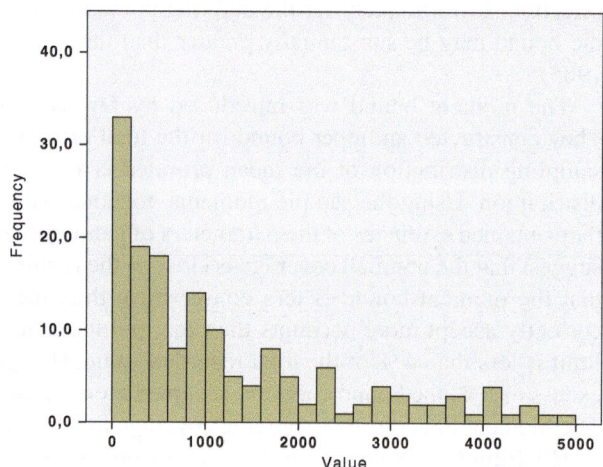

4   Simulations

In this section we will carry out a comparative analysis on the sampling designs
used in audit by means of a simulation study on distributions performed on a
real accounting population of credit invoices of an audited society. In Table 1 the
characteristics of the accounting population are presented (Fig. 1).

In the previous studies the distributions of recorded values were modelled by
skewed distributions (Smith 1979; Rohrbach 1993). In our study we shall adopt
simulations from the lognormal distribution with the estimated parameters by the
real accounting population, that is LogNormal(7.001, 1.71). On the other hand, as
the aim is to compare sampling plans in case of symmetric and non-symmetric
distribution, we have simulated from a Normal distribution with parameters $\mu =
5,530$ and $\sigma = 400$ estimated using the data of the real accounting. The population
size was set to $N = 3,000$. Following Horgan (1996), we simulated the percentage
of errors and the value of the overstatements. Five errors percentages were seeded on

the book values population: 1.8 %, 3.7 %, 5.5 %, 11.0 %, 16.5 % and the taint used values were 5 % and 10 % of the book value.

Moreover two hypothesis on the presence of the error on the items were considered: the first considers the error presence randomly within all the units of the account population, the second only on the units with book values greater than its third quartile thinking about a high concentration of big misstatements. In the end the study was based on 40 populations. Random samples of sizes $n = 60, 100, 200$ were drawn from the simulated populations. The replicates were $m = 1,000$. These sample sizes were chosen to reflect those used in audit practice and in the previous studies. The simple random SRS and the MUS without replacement were used in selecting samples units from the simulated study populations. The procedure of selection was the systematic procedure with random ordering of line items. The estimates of the error bound obtained from the simulated data were compared using the following performance criteria:

- Coverage probability: for a specific bound it refers to the proportion of replications for which a bound is greater than or equal to the true population error amount. A bound is considered unreliable if its coverage is significantly below the specified nominal coverage, otherwise it is reliable.
- Variability of the bound: this is an indicator of the uncertainty of the bound. It is measured by the mean squared error (MSE):

$$\frac{1}{1000} \sum_{i=1}^{1000} (\widehat{T_{z_i}} - T_z)^2 \tag{7}$$

where $\widehat{T_{z_i}}$ is the estimated value for the total error at the ith replicate.
- Relative Efficiency (D_{eff}): the relative efficiency is defined as the ratio of the MSE of the bounds of the MUS with respect to SRS.
- Mean: The mean of the bound is computed as follows:

$$AV = \frac{\sum_{i=1}^{1000} \widehat{T_{z_i}}}{1000}. \tag{8}$$

- Relative Advantage (RA): the RA is a way of comparing the relative tightness of the bounds of the MUS with respect to the SRS.

$$\frac{AV(MUS) - AV(SRS)}{AV(MUS)}. \tag{9}$$

When $0 \leq RA \leq 1$, the bound computed on the SRS is tighter than the MUS. The bound computed on the SRS is said to be more conservative than the MUS if RA < 0.

The discussion of the results consists of some general observations on the performance of the sampling designs. The performance of each sampling designs

Table 2 The simulation results at the 95 % confidence limit from Norm($\mu = 500$, $\sigma = 40$) of size $N=3,000$ with two error amounts of 5 and 10 % of the book values and five error rates on the population (error presence randomly on all the units)

n	Error rates (%)	Taint: 5 %			Taint: 10 %		
		SME	D_{eff}	RA	SME	D_{eff}	RA
60	1.8	0.001	0.989	−0.001	0.001	0.823	−0.103
100	1.8	0.001	0.981	−0.001	0.001	0.787	−0.165
200	1.8	0.001	1.003	−0.001	0.001	0.814	−0.299
60	3.7	0.002	0.793	−0.026	0.002	0.698	−0.124
100	3.7	0.001	0.865	−0.037	0.001	0.776	−0.190
200	3.7	0.001	0.846	−0.066	0.001	0.708	−0.340
60	5.5	0.002	0.754	−0.068	0.002	0.597	−0.166
100	5.5	0.001	0.723	−0.105	0.001	0.594	−0.254
200	5.5	0.001	0.763	−0.175	0.001	0.612	−0.429
60	11.0	0.003	0.717	−0.121	0.003	0.540	−0.212
100	11.0	0.002	0.705	−0.176	0.002	0.516	−0.321
200	11.0	0.001	0.706	−0.294	0.001	0.538	−0.518
60	16.5	0.003	0.630	−0.220	0.003	0.431	−0.316
100	16.5	0.002	0.663	−0.322	0.002	0.439	−0.461
200	16.5	0.001	0.617	−0.503	0.001	0.441	−0.702

Reporting the SME for MUS, the efficiency of MUS on SRS design over the Stringer bound (D_{eff}) and the relative advantage (RA) of MUS on SRS over the Stringer bound. Results are based on $m = 1,000$ replications

is evaluated on the distributions of book values (Normal/Lognormal distribution) with respect to the incidence of 100 % tainting (5 or 10 % of overstatement error), to the line item error rates (1.8 % which are 54 items with errors on $N=3,000$, 3.7 % which are 108 items with errors, 5.5 % which are 165 items with error, 11.0 % which are 330 items with errors and 16.5 % which are 495 items with errors) and to the hypothesis of the position of items with errors (H_1: randomly on all the units, H_2: on those units with book values greater than its third quartile). The discussion highlights the performance of the sampling designs with respect to the relative efficiency and the relative advantage of the estimators by different sampling designs.

4.1 The Case of the Normal Populations

Results are shown in Table 2 (for H_1) and Table 3 (for H_2). First of all, the coverage probabilities are always above the nominal level and equal to 100 % at the nominal 95 % for every sample in both the hypothesis. The column named D_{eff} is the relative efficiency of MUS over SRS.

From Table 2, $D_{eff} < 1$ in most cases implies that the MUS is as efficient as or more efficient than (lower MSE) the SRS. We can observe that the relative efficiency

Table 3 The simulation results at the 95 % confidence limit from Norm($\mu = 500$, $\sigma = 40$) of size $N = 3,000$ with two error amounts of 5 and 10 % of the book values and five error rates on the population (error presence on those units with book values greater that the third quartile)

n	Errors rates (%)	Taint: 5 %			Taint: 10 %		
		SME	D_{eff}	RA	SME	D_{eff}	RA
60	1.8	0.001	0.478	−0.001	0.001	0.510	−0.027
100	1.8	0.001	0.478	0.001	0.001	0.486	−0.053
200	1.8	0.001	0.438	0.003	0.001	0.495	−0.091
60	3.7	0.002	0.441	0.001	0.002	0.477	−0.056
100	3.7	0.001	0.429	0.001	0.001	0.515	−0.084
200	3.7	0.001	0.385	0.004	0.001	0.505	−0.152
60	5.5	0.002	0.461	−0.001	0.002	0.507	−0.083
100	5.5	0.001	0.445	0.006	0.001	0.485	−0.130
200	5.5	0.001	0.515	−0.004	0.001	0.514	−0.201
60	11.0	0.002	0.372	0.001	0.003	0.492	−0.147
100	11.0	0.002	0.425	−0.003	0.002	0.501	−0.198
200	11.0	0.001	0.490	0.004	0.001	0.518	−0.327
60	16.5	0.002	0.329	0.001	0.003	0.480	−0.190
100	16.5	0.002	0.369	0.002	0.002	0.525	−0.274
200	16.5	0.001	0.376	0.002	0.001	0.489	−0.413

Reporting the SME for MUS, the efficiency of MUS on SRS sampling design over the Stringer bound (D_{eff}) and the relative advantage (RA) of MUS on SRS over the Stringer bound. Results are based on $m = 1,000$ replications

for a given Error Rate increases with an increasing taint. This implies that the efficiency of the MUS increases with an increasing taint. The efficiency of the MUS is higher in populations with higher error rates. The relative efficiency also increases with the increase of a given taint. The relative advantage values are less than zero value RA < 0 of the MUS over the SRS in all cases, which means that the Stringer bound measured on the SRS sampling design is more conservative than the Stringer bound measured on the MUS sampling design, in other words the MUS is tighter than the SRS sampling design. Moreover the RA varies with the line item error rate, with the taint as well as with the sample size. In particular, the RA decreases with an increasing error rate. The higher efficiency occurs in populations with lower error rates. For the variation with taint, the RA is higher in audit populations containing a lower value of errors.

Table 3 reports the results for the hypothesis that the error presence is randomly on those units with book values greater than the third quartile. The relative efficiency of the MUS over the SRS was less than one in almost all cases. This implies that the MUS is more efficient than the SRS. In contrast with the hypothesis reported in Table 2, the relative efficiency for a given Error Rate decreases with an increasing taint. The relative advantage of the MUS over the SRS varies with the line item error rate, with the taint value as well as with the sample size. For the variation with taint model, the RA is almost close to zero for taint equals to 5 %. In this case the Stringer

Table 4 The simulation results at the 95 % confidence limit from LogNormal(7.001, 1.71) of size $N = 3,000$ with two error amounts of 5 and 10 % of the book values and five error rates on the population (error presence randomly on all the units)

		Taint: 5 %			Taint: 10 %		
n	Errors rates (%)	SME	D_{eff}	RA	SME	D_{eff}	RA
60	1.8	0.001	1.094	0.005	0.002	0.877	−0.013
100	1.8	0.001	0.923	0.012	0.002	0.735	−0.009
200	1.8	0.000	0.589	0.018	0.001	0.850	−0.016
60	3.7	0.001	0.824	−0.005	0.002	0.684	0.023
100	3.7	0.001	0.861	−0.005	0.002	0.692	0.011
200	3.7	0.001	0.877	0.002	0.001	0.633	−0.007
60	5.5	0.002	0.892	−0.003	0.004	0.890	−0.006
100	5.5	0.001	0.815	−0.008	0.003	0.855	−0.002
200	5.5	0.001	0.754	0.007	0.002	0.815	0.005
60	11.0	0.003	0.990	−0.013	0.004	0.788	−0.016
100	11.0	0.001	0.661	−0.016	0.003	0.866	−0.014
200	11.0	0.001	0.741	−0.020	0.002	0.856	−0.032
60	16.5	0.003	1.097	−0.044	0.005	0.948	−0.009
100	16.5	0.002	0.922	−0.065	0.003	0.731	−0.003
200	16.5	0.001	1.079	−0.080	0.002	0.821	0.007

Reporting the SME for MUS, the efficiency of MUS on SRS sampling design over the Stringer bound (D_{eff}) and the relative advantage (RA) of MUS on SRS over the Stringer bound. Results are based on $m = 1000$ replications

bound on the MUS was as conservative as the Stringer bound on SRS, indeed the RA values range from −0.3 to 0.6 % depending on the sample sizes. The RA is lower for an audit populations taint of 10 %. The RA ranges from −2.7 to −41.3 %, which means that the Stringer bound measured on the SRS sampling design is more conservative than the Stringer bound measured on the MUS sampling design, hence the MUS is tighter than the SRS. Given the Error Rates, the RA decreases with an increasing sample size, as the RA decreases with the increase of the Error Rates given the sample size.

4.2 The Case of the LogNormal Populations

Results are shown in Tables 4 (for H_1) and Table 5 (for H_2). First of all, the coverage probabilities are always above the nominal level and equal to 100 % at the nominal 95 % for every samples.

From Table 4, $D_{eff} > 1$ in most cases of the first taint value (5 %), which means that the MUS is less efficient (higher MSE values) than the SRS. On the other hand for the second value of taint values (10 %) the MUS is more efficient than the SRS. In the first case of taint values, the relative advantage values vary from −8 to 1.2 %.

Table 5 The simulation results at the 95 % confidence limit from LogNormal(7.001, 1.71) of size $N=3,000$ with two error amounts of 5 and 10 % of the book values and five error rates on the population (error presence on those units with book values greater that the third quartile)

n	Errors rates (%)	Taint: 5 %			Taint: 10 %		
		SME	D_{eff}	RA	SME	D_{eff}	RA
60	1.8	0.002	1.117	0.059	0.004	1.256	0.172
100	1.8	0.001	1.474	0.089	0.002	1.748	0.223
200	1.8	0.001	1.095	0.157	0.002	1.280	0.311
60	3.7	0.002	1.249	0.091	0.004	1.244	0.161
100	3.7	0.001	1.009	0.139	0.003	1.112	0.225
200	3.7	0.001	1.097	0.228	0.002	1.032	0.322
60	5.5	0.003	1.521	0.147	0.004	1.046	0.234
100	5.5	0.002	1.173	0.211	0.004	1.208	0.308
200	5.5	0.001	1.014	0.309	0.003	1.409	0.409
60	11.0	0.002	0.870	0.194	0.005	1.083	0.296
100	11.0	0.002	0.958	0.273	0.004	1.104	0.382
200	11.0	0.002	0.995	0.385	0.003	1.231	0.475
60	16.5	0.002	0.640	0.266	0.004	0.703	0.363
100	16.5	0.001	0.630	0.346	0.006	0.915	0.438
200	16.5	0.001	0.552	0.452	0.002	0.644	0.519

Reporting the SME for the MUS, the efficiency of the MUS on the SRS sampling design over the Stringer bound (D_{eff}) and the relative advantage (RA) of MUS on SRS over Stringer bound. Results are based on $m = 1000$ replications

Given the sample size, the RA values decrease when the error rates increase. In the case of taint values of 10 % there are only negative values of relative advantage, $RA < 0$, which means that the Stringer bound measured on the SRS sampling design is more conservative than Stringer bound measured on the MUS sampling design, in other words the MUS is tighter than the SRS sampling design. Moreover the RA varies with the line item error rate as well as the sample size. In particular, the RA decreases with an increasing error rate, given the sample size. The higher efficiency occurs in populations with lower error rates. Furthermore, given the Error Rates, the RA decreases with an increasing sample size.

Table 5 reports the results for the second hypothesis of the error presence. With regard to D_{eff}, the results do not show some of the regularities found in Table 4. In particular, when the taint value is 5 % the MUS is less efficient (higher MSE values) than the SRS for error rates from 1.8 to 5.5 %. On the other hand for the 10 % value taint the MUS is more efficient than the SRS only for an error rate equal to 16.5 %. The relative advantage always shows positive values which means that the Stringer bound measured on SRS sampling design is less conservative than Stringer bound measured on the MUS sampling design, in other words the SRS is tighter than the MUS sampling design. Moreover the RA varies with the line item error rate as well as with the sample size. In particular, the RA decreases with an increase of the error rate, given the sample size. Furthermore, given the error rate, the RA decreases with an increasing sample size. For a given error rate and sample size, the RA increases when the taint increases.

5 Conclusions

In this paper we offer a comparative study of simulations. In particular we have evaluated whether the sampling method in relation to the different shape distributions was able to distinguish between two hypothesis of the error presence on the book values: the first regards random error on book values, the second regards the "systematic" error only on the higher book values. We have distinguished between an unreliable symmetric distribution and a realistic non-symmetric distribution for the book values population. We have measured the Stringer bound which in literature has already been regarded as very conservative. Some interesting considerations have arisen. For the Normal distribution case, in the hypothesis of random presence of the errors, the MUS is almost more efficient (lower MSE) than a SRS sampling design and the Stringer bound measured on the MUS sampling design proves to be tighter than the SRS. On the other hand in the hypothesis of "systematic" error presence on the higher book values even if the MUS is always more efficient than the SRS, only when the taint value is greater, the Stringer bound measured on the MUS sampling design is always tighter than the SRS sampling design. Obviously this increase in efficiency is due to the conservativeness of the Stringer bound that becomes more discriminant when the sampling plan is more sophisticated in case the book values come from a Normal distribution. The case of the Lognormal distribution shows differences regarding the hypothesis of the error presence: in fact the MUS proves to be more efficient and tighter than the SRS only in the case of the random error hypothesis. These conclusions indicate that it is fundamental for auditors to verify at least which distribution follows the total book amount in order to choose the most appropriate sampling plan for estimating the total error amount. Future developments of this work include the evaluation of other different bounds on a wide range of distributions including different shapes.

References

American Institute of Certified Public Accountants (AICPA): Audit Guide: Audit Sampling. AICPA, New York (2008)

Anderson, R.J., Teitlebaum, A.D.: Dollar-unit Sampling. Canadian Chartered Accountant, pp. 30–39 (April, 1973)

Arens, A.A., Loebbecke, J.K.: Applications of Statistical Sampling to Auditing. Prentice-Hall, Englewood Cliffs (1981)

Beck, P.J.: A critical analysis of the regression estimator in audit sampling. J. Account. Res. **18**, 16–37 (1980)

Bickel, P.J.: Inference and auditing: The stringer bound. Int. Stat. Rev. **60**(2), 197–209 (1992)

Dworin, L., Grimlund, R.A.: Dollar unit sampling for accounts receivables and inventory. Account. Rev. **59**, 218–241 (1984)

Dworin, L., Grimlund, R.A.: Dollar-unit sampling: A comparison of the quasi-Bayesian and moment bounds. Account. Rev. **61**(1), 36–58 (1986)

Felix, W.L., Jr., Leslie, D.A. Neter, J.: University of Georgia Center for Audit Research Monetary-Unit Sampling Conference, 24 March 1981. Auditing J. Pract. Theory **1**(2), 92–103 (1981)

Fienberg, S.E, Neter, J., Leitch, R.A.: Estimating the total overstatement error in accounting population. J. Am. Stat. Assoc. **72**(358), 295–302 (1977)

Financial Accounting Standards Board (FASB): Statement of Financial Accounting Concepts No. 2 (SFAC2 - Qualitative Characteristics of Accounting Information). Stamford, Conn.: FASB (1980)

Fishman, G.S.: Confidence intervals for the mean in the bounded case. Stat. Probab. Lett. **12**, 223–227 (1991)

Goodfellow, J.L., Loebbecke, J.K., Neter, J.: Some perspectives on CAV sampling plans. Part I *CA* Magazine, October, pp. 23–30; Part II *CA* Magazine, November, pp. 46–53 (1974)

Grimlund, R.A., Felix, D.W., Jr.: Simulation evidence and analysis of alternative methods of evaluating dollar-unit samples. Account. Rev. **62**(3), 455–480 (1987)

Ham, J., Losell, D., Smieliauskas, W.: An empirical study of error characteristics in accounting populations. Account. Rev. **60**(3), 387–406 (1985)

Hansen, M.H., Hurwitz, W.N.: On the theory of sampling from finite populations. Ann. Math. Stat. **14**, 332–362 (1943)

Horgan, J.M.: The moment bound with unrestricted random, cell and sieve sampling of monetary units. J. Account. Bus. Res. **26**(3), 215–223 (1996)

Horvitz, D.G., Thompson, D.J.: A generalization of sampling without replacement from a finite universe. J. Am. Stat. Assoc. **47**, 663–685 (1952)

Howard, R.C.: A combined bound for errors in auditing based on Hoeffding's inequality and bootstrap. J. Bus. Econ. Stat. Am. Stat. Assoc. **12**(4), 437–448 (1994)

Johnson, J.R., Leitch, R.A., Neter, J.: Characteristics of errors in accounts receivables and inventory audits. Account. Rev. **58**, 270–293 (1981)

Kaplan, R.S.: Statistical sampling in auditing with auxiliary information estimators. J. Account. Res. **11**(2), 238–258 (1973)

Kaplan, R.S.: Sample size computations for dollar-unit sampling. J. Account. Res. **13**(3), 126–133 (1975)

Kvanli, A.H., Shen, Y.K., Deng, L.Y.: Construction of confidence intervals for the mean of a population containing many zero values. J. Bus. Econ. Stat. Am. Stat. Assoc. **16**(3), 362–368 (1998)

Leitch, R.A., Neter, J., Plante, R., Sinha, P.: Modified multinomial bounds for larger number of errors in audits. Account. Rev. **57**(2), 384–400 (1982)

Leslie, D.A., Teitlebaum, A.D., Anderson, R.J.: Dollar-Unit Sampling - A Practical Guide for Auditors. Pitman, London (1979)

Modow, W.G.: On the theory of systematic sampling. Ann. Math. Stat. **20**, 333–354 (1949)

Neter, J., Loebbecke, J.K.: Behaviour of Major Statistical Estimators in Sampling Accounting Population - An Empirical Study. AICP, New York (1975)

Neter, J., Loebbecke, J.K.: On the behavior of statistical estimators when sampling accounting populations. J. Am. Stat. Assoc. **72**(359), 501–507 (1977)

Neter, J., Johnson, J.R., Leitch, R.A.: Characteristics of dollar-unit taints and error rates in accounts receivables and inventory. Account. Rev. **60**, 488–499 (1985)

Pap, G., van Zuijlen, M.C.A.: On the asymptotic behaviour of the stringer bound. Stat. Neerl. **50**(3), 367–389 (1996)

Plante, R., Neter, J., Leitch, R.A.: Comparative performance of the multinomial, cell and stringer bound. Auditing J. Pract. Theory **5**, 40–56 (1985)

Ramage, J.G., Kreieger, A.M., Spero, L.L.: An empirical study of error characteristics in audit populations. J. Account. Res. **17**(Suppl.), 72–102 (1979)

Reneau, J.H.: CAV bounds in dollar unit sampling: some simulation results. Account. Rev. **53**(3), 669–680 (1978)

Rohrbach, K.J.: Variance augmentation to achieve nominal coverage probability in sampling from audit populations. Auditing J. Pract. Theory **12**(2), 79–97 (1993)

Smieliauskas, W.: Control of sampling risks in auditing. Contemp. Account. Res. **3**(1), 102–124 (1986)

Smith, T.M.F.: Statistical sampling in: Auditing: A Statistician's viewpoint. Statistician **28**(4), 267–280 (1979)

Stringer, K.W.: Practical aspects of statistical sampling in auditing. In: Proceedings of the Business and Economics Statistics Section, pp. 405–411. ASA (1963)

Swinamer, K., Lesperance, M.L., Will, H.: Optimal bounds used in dollar unit sampling: A comparison of reliability and efficiency. Communications in Statistics - Simulation and Computation, **33** (1), pp. 109–143 (2004)

Tsui, K.W., Matsumura, E.M., Tsui, K.L.: Multinomial-Dirichlet bounds for dollar-unit sampling in auditing. Account. Rev. **60**(1), 76–97 (1985)

Wurst, J., Neter, J., Godfrey, J.: Comparison of sieve sampling with random and cell sampling of monetary units. Statistician **11**(2), 235–249 (1989)

Implementing the First ISTAT Survey of Homeless Population by Indirect Sampling and Weight Sharing Method

Claudia De Vitiis, Stefano Falorsi, and Francesca Inglese

Abstract The Italian National Institute of Statistics carried out the first survey on the homeless population. The survey aims at estimating the unknown size and some demographic and social characteristics of this population. The sample strategy for the homeless survey refers to the theory of indirect sampling, based on the use of a sampling frame indirectly related to the target population. In fact, the methodological strategy used to investigate the homeless population could not follow the standard approaches of the official statistics usually based on the use of population lists. In the general context of hard-to-reach population methods, the survey used a sample approach based on a time-location sampling, in which target population units are reached through the random selection of places (the centers) and instants of time. The availability of information about centers providing services to people living in extreme poverty was the fundamental information for the design of the survey. Following the indirect sampling approach, the estimation is performed through the "weight-sharing method," based on the links connecting the frame of services with the population of the homeless.

Keywords Hard-to-reach population • Time-location sampling

C. De Vitiis (✉) • S. Falorsi • F. Inglese
Istituto Nazionale di Statistica, via Tuscolana 1788, 00184 Roma, Italy
e-mail: devitiis@istat.it; stfalors@istat.it; fringles@istat.it

F. Mecatti et al. (eds.), *Contributions to Sampling Statistics*, Contributions to Statistics, DOI 10.1007/978-3-319-05320-2__8,

119

1 Introduction

The survey on homeless people, conducted for the first time in Italy in 2011–2012, is an important component of a research project on the condition of people living in extreme poverty. The research project,[1] launched in 2009, was aimed at the dual purpose of building an archive of the system of formal and informal services, public and private, existing in the country aimed to meet the needs of people living in extreme poverty and studying the phenomenon of homeless people spread over the Italian territory. To achieve these objectives three separate and successive surveys were carried out with the following purposes: (1) the construction of an archive relative to the population of centers (organizations/entities) providing services to people living in extreme poverty (2009–2010), (2) the acquisition of detailed information on the services provided by each one of the listed centers (2010–2011), (3) the realization of the sample survey on homeless people (2011–2012).

The population of homeless people is composed of all individuals who are experiencing housing problems due to the impossibility and/or inability to obtain independently and maintain a home in the true sense (*ETHOS typology*[2]). In this population people living in public spaces (streets, shacks, abandoned cars, caravans, sheds), in night dormitories, in hostels for homeless people or temporary housing, in housing for specific social support interventions (for homeless individuals, couples and groups) are included.

The availability of information about centers providing services to people living in extreme poverty was the fundamental information for the design of the survey. Because a direct list of individuals belonging to the target population was not available, the sampling strategy could not be based on a classic direct sampling approach.

The focus of this paper is on the application of the theoretical framework of indirect sampling and weight-sharing method to the context of the Italian official statistics; in particular the application with regard to a very peculiar survey on a population never investigated before at national scale and required, therefore, a very refined implementation plan in order to respect all the theoretical hypothesis.

In general, the planning of a survey on homeless people has to deal with several difficulties, both in terms of data collection and methodology, due primarily to the fact that a list is not available to identify directly the units belonging to this population. Investigating this particular population requires, therefore, reaching statistical units in a different way, for example, by individuating centers where they regularly go to receive the services they need.

This situation can be dealt with in the general context of *hard-to-reach* population methods (Marpsat and Razafindratsima 2010), in particular by means of sample

---

[1]The Ministries of Health and Labour and Social Policies, Istat, the Italian Federation of associations for the Homeless (fio.PSD) and Caritas took part in the project.

[2]Source: http://www.feantsa.org/files/indicators_wg/ETHOS2006/EN_EthosLeaflet.pdf.

designs classified as *location sampling* or *time-location sampling* (Kalton 2001), in which target population units are reached through the random selection of places (the centers) and instants of time. One of the first experiences of sampling in the field of the homeless population restricted to a limited territory such as a city was carried out in Washington in 1991 (Dennis and Iachan 1993), based on a multiframe approach.

The methodology adopted for the nationwide Italian survey finds actually its theoretical basis in the *indirect sampling* (Lavallée 2007) which is founded on the idea of using a sampling frame referred to a population that is linked indirectly to the target population. In this approach, therefore, the sampling design is defined on the basis of information contained in a list that refers to a different population from that of interest.

The centers chosen for the survey are those providing night shelter and canteen services: the former are by definition, services frequented exclusively by homeless people, while the latter are services frequented not only by homeless people but also by people who are living in a state of extreme poverty. Thus, the sampling list is represented by provisions, meals and beds, provided by the services. In this perspective, the population of interest is restricted to homeless people who frequent services of night shelters and canteen services, excluding people receiving services from different types of centers, where conducting interviews on persons would imply different survey techniques.

The method used to estimate the parameters of the population of interest is the *weight-sharing method* which is based on the knowledge of the links between units belonging to the two populations. The links between the sample units, i.e., provisions, and survey units, homeless people, are one-to-one, in the sense that each service provided can be associated, at a given instant of time, only to a homeless person. The *weight-sharing method* allows to deal with the duplication issue arising from the fact that individuals may make multiple visits during the survey's time frame, the increased selection probabilities associated with multiple visits need to be taken into account in developing the survey weights.

To ensure a correct sharing of the weights, the map of the links between persons and centers has to be known: for each interviewed person the list of all visited centers had to be collected for a fixed period of time. The length of the observation period for each person has been set in a week, so that the estimate of the number of links refers to an average week. The survey instrument to collect appropriately all the information to map the links was a daily diary, in which the exact places where people ate and slept in the 7 days preceding the interview were collected, individuating all centers belonging to the considered list.

The methodology used for this study presents some limitations: the first is related to the coverage of the phenomenon, because part of the population of interest, consisting in homeless people living in public spaces and who do not use services or night shelters or canteen services, is not surveyed; the second is due to the fact that the services, particularly the canteen, are also frequented by people who do not belong to the population of interest. Both problems can introduce bias and/or increased variability (in case of over-coverage) in the estimate of the population of interest, the first implying a problem of under-coverage of the phenomenon and the second a problem of over-coverage due to the fact that the population caught at the centers does not reflect the true target population.

The paper is organized as follows. In Sect. 2 the preliminary survey steps are presented to describe the construction of the archive of centers. Section 3 describes the sampling design, while Sect. 4 illustrates the estimation procedure. Finally, Sect. 5 reports the main results of the survey about homeless people and Sect. 6 some concluding remarks.

2   The First Operational Phases for the Construction of the Archive of Centers

The homelessness phenomenon is mainly spread in the wider cities, where the presence of people in difficulty and services provided to them is more relevant; therefore the survey was limited on a subset of Italian municipalities, selected on the basis of their demographic size and beforehand. These are 158 municipalities, including all the municipalities with over 70,000 inhabitants, the provincial capitals with more than 30,000 inhabitants, and the municipalities bordering on the municipality with more than 250,000 inhabitants. The census of service providers for homeless people was conducted in these towns with special attention to soup kitchens and night shelters that represent the places with the highest probability of finding homeless people.

The archive of the centers has been obtained through two phases: (a) a census of the organization (centers) offering services to homeless people; (b) an in-depth survey on the service providers in order to collect information, both quantitative and qualitative, about their users (Pannnuzi et al, 2009).

The purpose was to build a database which, for each detected service, contains all the necessary information: service typology, service details (number of bed spaces, average number of meals provided per day, average number of clients per day), supplier organization denomination, address, phone, possible organizations on behalf of the service is provided, organization representative, and organization type.

This information is being obtained by a CATI survey. Starting from the information contained in the pre-existing Istat, Caritas and fio.PSD databases, the survey updated and completed the picture by adding new organizations, reported by the already interviewed organizations. The added organizations are being interviewed in the same way, with a snowball technique in order to catch the maximum number of centers, even informal, supplying services to the homeless.

Once the database of the centers was complete, the services and, hence, the organizations were surveyed by a CAPI interview. A deep reference frame on the situation of the active services and organizations on the territory for the homeless has been drawn.

The information mainly regarded: the basic organizational and service details (contact details and location); the type of organization (whether municipal or other public bodies, private, NGO, etc.); the geographical area served; the user's main characteristics (age, gender, citizenship, household type, presence of any physical or mental restrictions); the service access criteria; the provision of any support to exit from the homelessness; the collection of data by the organization or the service; the funding sources and the share of resources for the homeless; the staff information; the cooperation among the services and the interactions with other organizations, especially with social and health units; the participation of the organization in workshops, seminars about the homeless problems; the client participation in the organizational activities.

The information collected at the centers about daily services, number of total and homeless users, opening days, represented the key information for the definition of the sampling reference universe for the survey on homeless people.

3 The Sampling Design for the Survey on Homeless People

3.1 Objectives of the Survey and Reference Population

The main objective of the survey was to estimate—at national level—the unknown size of the target population, along with some demographic and social characteristics of homeless people. The survey was planned with the aim of measuring the extent of a phenomenon never observed in our country, if not in local contexts, and to obtain information on the process underlying the causes of social exclusion of homeless people.

The observation field of the survey was delimited within the geographical scope of the 158 municipalities, identified on the basis of population size, and all the centers (more than 800) in which night shelter and canteen services are provided, listed in the archive of centers. The services of night shelters are self-managed housing, shelters, dormitories, dormitories emergency, residential communities and semi-residential community. The canteen services include both those that provide a meal for lunch and those who provide a meal for dinner.

From the peculiarities of the phenomenon under observation a need arose to use a definition of the population connected to the moment in which the condition of homelessness has occurred and to define an observational period delimited in a precise interval of time. The homeless are, in fact, a population subject to constant changes that may be caused by different phenomena. It is a population that is renewed over time and from one period to another the size and composition of the population may change due to demographic shifts or as an effect of the evolution of society in which, for example, the working condition is strongly precarious.

In order to identify correctly the homeless people at the survey stage, it was adopted as a reference period of the occurrence of the event that led an individual to sleep in centers that provide night shelters or on the street, the month before with respect to the day of the interview.

Furthermore, as the size of the population of homeless people strongly depends on the period in which the phenomenon is observed, the period of observation has been established in 30 days, a period considered long enough to ensure that the majority of homeless people used at least once the services involved in the investigation. The choice of the reference period of the survey was determined by considerations related to the organization of the network for the data collection and the evaluation of experts who felt that the part of population missed in a winter month was of a limited scale. For these reasons, the period of time fixed for the survey of the phenomenon was fixed as the 30 days between November 21st and December 20th 2011.

The delimitation of the observation field and reference period for the survey led to a more stringent definition of the target population. In fact, it consists of all the homeless people who receive at least one provision provided at one of the centers, belonging to the set C, that provide night shelters or canteen services during the fixed period of time J in the 158 municipalities constituting the territorial scope of reference.

3.2   Sampling Design

The sampling design was defined with regard to the population of provisions $U^A(J)$, with reference to the period J. This set of provisions constitutes the reference universe for the survey, i.e., the population $U^A(J)$ of size $N^A(J)$ used for reaching the target population $U^B(J)$ of size $N^B(J)$.

Each provision is individuated by the triplet (center-day-provision), as it will be clarified below. The sampling frame was composed starting from the list of all the centers—operating in the defined territory—that provide night shelter and canteen services, collected during the first operational phases of the project. The centers are the places where services are provided and a center may provide multiple services.

To each center c ($c = 1, \ldots, C$) the opening days in a month were associated, in order to build the center-day couple constituting the statistical units of the list. For the definition of the sampling list the services involved in the survey were classified into three types: nightly accommodation, canteens providing meals for lunch, and canteens providing meals for dinner. The reason of this definition is that the services thus defined are associated to each center so as to be identified by a unique center code, ensuring the univocal correspondence between provisions and individual, as each individual can receive only one provision in each center during a single time interval g ($g = 1, \ldots, G$), called "day."

The sampling design adopted for the survey is a two-stage stratified random sampling in which all centers are included and each of them represents a separate stratum of provisions. This choice was made because the number of such centers allowed visiting them all, and due to efficiency considerations of the design, since in this way a stage of selection was avoided.

The first-stage units are the opening days of the centers C in the reference period of the survey J, each corresponding to a cluster of provisions (meals or beds). The final statistical units (or secondary statistical units) are represented by the provisions (each of them connected through a one-to one relationship to one individual) delivered in opening-days defined in the list.

In order to ensure the coverage of all the days of the reference month, the allocation of the sample days to centers was done randomly, from the list of opening days of each service.

The aim of the sample scheme was to achieve a sample of provisions selected with equal probabilities, since the goal of the survey was to produce estimates at national level. Moreover, information about the interest variables that could guide the definition of a sample design of a different type (for example, with nonproportional allocation or selection with unequal probabilities) was not available.

3.3   Sample Size and Allocation

The sample size was determined in relation to the monthly provisions to homeless people in the considered centers. The total amount of monthly provisions was estimated, using the information on the services contained in the archive of the centers, by considering the average number of daily provisions for each center and the number of opening days of services in a month; the estimate of monthly provisions provided to homeless has been obtained by applying to the estimate of monthly provisions, the percentage of homeless users.

The sample size was fixed at about 5,400 units after an assessment based on the calculation of the expected sampling error of the estimate of the population size at national level, which was the planned territorial domain.

This evaluation was performed, due to the unavailability of prior information about the variability of the individual attendance of services, on the basis of a likely and simplified conjecture on the variability of the individual number of links, assuming that the variability of the number of links is one of the main sources of variability for the estimates, together with the time variability. The expected sampling coefficient of variation for the estimate of the population size was around 3.5 %.

In order to achieve a self-weighting sample, the sample size to be assigned to each center has been calculated on the basis of a fixed sample fraction defined as

$$f^A(J) = \frac{n^A(J)}{\tilde{N}^A(J)},\qquad(1)$$

where $n^A(J)$ indicates the overall sample size and $\tilde{N}^A(J)$ the total estimated monthly number of provisions provided to the population of interest in the period J. Having thus defined the proportional allocation of the sample among the centers, the total number of provisions attributed to each service is closely related to the size of the center. The sample size in each center c is thus determined as

$$n_c^A(J) = \tilde{N}_c^A(J) f^A(J). \tag{2}$$

3.4 Assigning the Number of Sample Days to Centers

After assigning to each center the total number of sample provisions, $n_c^A(J)$, to be sampled in the reference month of the survey, the number of provisions for each day of investigation g and the number of sample days to be attributed to each center were defined on the basis of fixed rules.

The number of provisions to be selected for every survey occasion has been calculated taking into account the maximum number of interviews that each interviewer would be able to conduct in the selected center-day during opening hours. This number has been fixed at four interviews per day per interviewer. For small centers a number of interviews ranging from a minimum of one to a maximum of four was attributed; centers of medium size were assigned a number of interviews equal to four; to larger centers, the number of interviews attributed goes from a minimum of 4 to a maximum of 12 (carried out by one to three interviewers).

The number of sample days to centers was assigned by fixing that the maximum number of occasions did not exceed 15 days. In this way two different needs were met: on the one hand to seize variability of the phenomenon through the flow of people in services during the reference period of the survey, and on the other hand to limit the presence of interviewers in services throughout the reference period of the survey.

3.5 Selection of Provisions

For each selected primary unit, defined by the couple center-day, users of service were selected at random by the interviewers each time on the spot. The selection was carried out, following a systematic procedure, from the list of the users or from the waiting incoming line, allowing the replacement of the persons selected that do not belong to the population of homeless people.

The interviewer selected individuals to be interviewed at random using a sampling interval defined on the basis of two quantities: the number of people to be interviewed and the number of daily predicted provisions for the specific center, obtained from the second phase survey. Strictly speaking, the sampling interval should be calculated on the actual number of users present on the day of the interview, but the interviewer, especially in the canteens where the flow is not predictable, did not know this number in advance and had therefore to use "expected" information.

However, in order to calculate afterwards the exact probability of inclusion of selected provisions, at the end of each survey day the interviewer took note of the effective number of users of the service on that occasion.

3.6 Inclusion Probabilities

The inclusion probabilities of selected provisions were calculated according to the implemented selection procedure. The inclusion probability of the primary units, identified by the couple of the generic g sample day in the center c (during the period J), is defined as

$$\pi_{cg}^A = \frac{d_c(J)}{D_c(J)}, \tag{3}$$

where $d_c(J)$ is the number of sample days of the center c and $D_c(J)$ is the number of opening days of the center c in the reference period J.

The second-stage probability of inclusion of the selected provisions i in the center c conditional on the selection of the day g is given by

$$\pi_{i|cg}^A = \frac{n_{cg}^A}{\tilde{N}_{cg}^A}, \tag{4}$$

where n_{cg}^A is the number of provisions on the day selected in the center cg and $\tilde{N}_{cg}^A$ is the total number of effective provisions provided to homeless people in the center-day cg. Finally, the probability of inclusion of the provision i in the center-day cg is obtained as

$$\pi_{cgi}^A = \pi_{cg}^A \, \pi_{i|cg}^A. \tag{5}$$

4 Estimation Procedure

4.1 Outline

Using *indirect sampling*, for every unit cgi selected in the sample s^A from $U^A(J)$ (provisions), the unit k of target population $U^B(J)$ (homeless persons), which has a correspondence with the unit cgi, (i.e., $l_{(cg)ik}^{AB} = 1$), is identified. In this way the sample s^B of units belonging to $U^B(J)$, for which it is possible to obtain interest estimates, is defined as

$$s^B = \left\{ k \in U^B(J) \,|\, \exists cgi \in s^A, \; l^{AB}_{(cg)ik} = 1 \right\}. \tag{6}$$

The two linked populations consist the first of the provisions provided in the set of center-days in the period J, $U^A(J) = \underset{cg}{\cup} U^A_{cg}(J)$, and the second, of all the homeless people who visit the set of center-days in the period J, $U^B(J) = \underset{cg}{\cup} U^B_{cg}(J)$.

In this methodological context, the method used for the calculation of weights to be given to individual units is the *weight-sharing method* or *generalized weight-sharing method* (GWSM, Deville and Lavallée 2006).

The method allows the reconstruction of the weight of the units of the sample s^B starting from the sampling weight associated to the units selected in the sample s^A. It takes into account the inclusion probability, π^A_{cgi}, of the selected sample units through a calculation based on the relationship between the units listed in the sampling frame $U^A(J)$ and the units belonging to the population of interest $U^B(J)$. For each interviewed unit k belonging to $U^B(J)$ the sampling weight is given by

$$w^B_k = 1/L^B_k(J) \sum_{s^A} w^A_{cgi} \, l^{AB}_{(cg)\,ik}, \tag{7}$$

where: w^A_{cgi} is the weight associated with the unit center-day-provision (cgi) selected in the sample s^A with inclusion probability π^A_{cgi} and $L^B_k(J) = \sum_{i \in U^A(J)} l^{AB}_{(cg)ik}$ is the total number of links of the individual k with all the provisions of $U^A(J)$ received during the period J in centers belonging to C (Ardilly and Le Blanc 2001a).

The correspondence between the two populations $U^A(J)$ and $U^B(J)$ can be represented by a *link matrix* $\Theta_{AB} = [l^{AB}_{(cg)\,ik}]$ of size $N^A(J) \times N^B(J)$, where each element $l^{AB}_{(cg)\,ik} \geq 0$. The existence of at least one link between all units of the target population $U^B(J)$ and the units of the population $U^A(J)$ is a necessary condition to ensure the unbiasedness of the estimator based on the weight-sharing method (Lavallée 2001).

Theoretically it would be necessary to map the links between the two populations on the entire reference period of the survey J. Generally, individuals subjected to interview are not able to provide information on their attendance of the centers for a period of time too long nor can know in advance which centers they would attend after the date of the interview. For these reasons the links between the two populations were observed over a period of time limited in a week as this was considered an observation period in which the behavior of a person about the attendance of the centers can be considered regular compared, for example, to that observed in a single day.

The collection of information over a period of a week allows limiting non-response in the diary, more than the observation of a longer period of time. The reference week is, therefore, not a fixed week but for each individual, is a sliding group of 7 days.

4.2   Parameters and Estimation

The main parameters of interest in the target population—defined in the survey reference period J—are total of "fixed" variables, i.e., variables that do not vary over time J (age, sex, nationality, etc.).

An important parameter of interest is represented by the unknown size of the population $U^B(J)$.

The total of a generic variable of interest y^B, considered as fixed during the period J, is

$$Y^B(J) = \sum_{k \in U^B(J)} y_k^B, \tag{8}$$

while the unknown size $N^B(J)$ of the target population $U^B(J)$ is obtained by setting

$$y_k^B = 1, \quad \forall k \in U^B(J). \tag{9}$$

Let us define the application K (Ardilly and Le Blanc 2001a) that relates all the provisions served in the period J in the set of centers C to each individual who has received it:

- K: (provisions) $\rightarrow$ (individual).
- $i = K(i)$.

The parameter of interest $Y^B(J)$, therefore, can be rewritten in the form

$$Y^B(J) = \sum_{k \in U^B(J)} y_k^B = \sum_{i \in U^A(J)} \frac{y_{K(i)}^B}{L_{K\,(i)}^B(J)}, \tag{10}$$

based on the assumption that the generic variable of interest y^B takes the same value for all the provisions related to the individual k, that is $K(i) = k$.

The total $Y^B(J)$ can be estimated on the population of provisions $U^A(J)$ through the transformed variable z_i^A, defined on the population of provisions $U^A(J)$,

$$z_i^A = \sum_{k \in U^B(J)} \frac{l_{ik}^{AB}}{L_k^B(J)} y_k^B \tag{11}$$

from which the equality between the totals of the two variables z_i^A and y_k^B originates

$$Z^A(J) = \sum_{i \in U^A(J)} z_i^A = \sum_{k \in U^B(J)} y_k^B = Y^B(J). \tag{12}$$

The total $Y^B(J) = Z^A(J)$ can be estimated on variable z_i^A as

$$\widehat{Z}^A(J) = \sum_{i \in U^A(J)} w_i^A \, t_i^A \, z_i^A = \widehat{Y}^B(J), \tag{13}$$

in which $t_i^A = 1$ if unit $i \in s^A$ and $t_i^A = 0$ if unit $i \notin s^A$.

By substituting (11) in (13) it is possible to obtain

$$\widehat{Z}^A(J) = \sum_{i \in U^A(J)} w_i^A \, t_i^A \sum_{k \in U^B(J)} \frac{l_{ik}^{AB}}{L_k^B(J)} y_k^B = \widehat{Y}^B(J), \tag{14}$$

where the second summation refers to sample s^B and therefore the estimator of $Y^B(J)$ assumes the form

$$\widehat{Y}^B(J) = \sum_{s^B} y_k^B \left(\sum_{s^A} w_i^A \frac{l_{ik}^{AB}}{L_k^B(J)} \right) = \sum_{s^B} w_k^B \, y_k^B, \tag{15}$$

in which, for simplicity, it is omitted to express the provision index in the full form (*cgi*) relative to the two stages of sampling.

Properties of GWSM about unbiasedness and variance evaluation using usual Horvitz–Thompson estimators are described and proved by Deville and Lavallée (2006).

Since in practice a weekly weight is given to each person, as it is based on the links referred to a week, the estimate of interest parameters related to the whole reference period, J, was only obtainable through a strong hypothesis about stability of the behavior of individuals among weeks, each of which can, therefore, be considered as representative of the whole period.

Under this hypothesis it is possible to expand the links observed for a week to the total period J by simply multiplying the number of weekly links by the number of survey weeks (Ardilly and Le Blanc 2001b).

In conclusion, the final weight for individual k is expressed as

$$w_k^* = \frac{1}{HL_k^B(\eta)} \sum_{s^A} w_i^A \, l_{ik}^{AB}, \tag{16}$$

where:

- $L^B_{K(i)}(\eta) = \sum\limits_{i \in U^A(\eta)} l^{AB}_{ik}$.
- η is the generic week.
- H is the number of weeks in month J.
- $U^A(\eta)$ is the part of population $U^A(J)$ restricted to week η.

4.3 Weighting Computation

In the estimation phase, for the application of the weight-sharing method, some preliminary operations were required to calculate the weights of the provisions in the sample s^A and to construct the link matrix, expressing the correspondence between units belonging to the population $U^B(J)$ and the units belonging to the population $U^A(J)$.

To compute properly the weights of provisions, the inclusion probability of each provision for each sample center-day (cgi) were calculated and correction factors have been introduced to take account of total non-response, determined by the non-response of centers and survey days. In fact, during the survey field individuals who refused to cooperate were replaced; this last form of non-response was not treated in the estimation phase.

Besides the classic form of non-recording of information on sampling units, in the homeless survey non-response was determined also by failure to collect essential information for the reconstruction of the links between people and provisions, contained in the diary section of the questionnaire.

This form of partial non-response, called link non-response, is associated with the situation where it is not possible to establish whether a unit k from $U^B(J)$ is linked to a unit $(cg)i$ from $U^A(J)$. This type of non-response constitutes a problem for the identification of the link (Xiaojian and Lavallée 2009).

An incorrect reconstruction of the links between the two populations involves the risk of introducing biases in the estimation of the parameters of interest, caused by incorrect counts.

The construction of the link matrix implies, in the presence of link non-response, the treatment of missing links. It is a necessary operation to associate to each individual k the weights of the provisions w^A_{cgi} received in the sample center-day cg and to determine the total number of links of the individual k with the population of provisions, $L^B_k(J)$.

4.3.1 The Sampling Weight of Selected Provisions

The first operation carried out for the construction of the weights of provisions involved the calculus of the inclusion probability of provisions provided to homeless people selected in the sample s^A, according to formulas in Sect. 3.6. To this aim some pieces of information have been used, obtained during the survey, regarding the number of provisions provided at each center-day sample cg, and collected in the second phase, regarding the information on the homeless user in centers.

The importance of combining these two pieces of information is determined by the fact that the canteen services are also frequented by individuals who are not part of the population of interest. Therefore, the number $\tilde{N}_{cg}^A$ of actual provisions delivered to homeless people in sample center-days was calculated by multiplying the number of total provisions of the specific day by the proportion of homeless users of the center.

It is worth underlining that the high variability of the influx to the centers does not guarantee that the probability of inclusion of all statistical units in the sample is invariable, as it was planned in the theoretical sample design. In fact, as the number of interviews per day is fixed, the probability may actually vary within selected days for the same center and among centers. The inclusion probabilities, anyway, were correctly calculated after the survey, even if they resulted affected by a light variability, not complying precisely the self-weighted theoretical sampling design.

4.3.2 Total Non-response Adjustment

With regard to the lack of participation of centers, which in fact constitutes a non-response of strata, a post-stratification has been applied on direct weights, in order to ensure that the final weights sum to the known total number of provisions provided to the homeless population in a generic month.

In fact, in the context of indirect sampling, where no known totals on the target population are available, the size of the reference population consists of the total number of monthly provisions provided in the centers considered, with reference to the period of observation.

The post-strata have been defined through the analysis of data. For centers that provide night shelter the geographical distribution of the centers was significant, together with the size of the centers in terms of provisions per month, while for canteen services only the size of the centers resulted in affecting response, thus defining the post-strata as classes by size in terms of provisions.

For each class the post-correction factor was calculated as the ratio between the total archive number of monthly homeless provisions (the known total) $_{ps}\tilde{N}^A(J)$ and the number of monthly homeless provisions of respondent centers $_{ps}N_R^A(J)$,

$$_{ps}f = \frac{_{ps}\tilde{N}^A(J)}{_{ps}N_R^A(J)}. \tag{17}$$

Table 1 Distribution of centers (total and respondents) and of homeless people interviewed by geographical area and type of center

Area	Type of center	Number of centers		Number of interviews
		Total	Respondent	
North	Meals	161	133	1,477
	Overnight accommodation	305	280	1,176
	Total	466	413	2,653
Center	Meals	69	55	540
	Overnight accommodation	96	78	377
	Total	165	133	917
South	Meals	98	82	664
	Overnight accommodation	73	63	462
	Total	171	145	1,126
Italy	Meals	328	270	2,681
	Overnight accommodation	474	421	2,015
	Total	802	691	4,696

The correction for first-stage unit non-response of the sampled days was carried out by the calculation of a correction factor for each center, γ_{cg}, considering the effective number of days of interview $d_c^*(J)$,

$$\gamma_{cg} = \frac{d_c(J)}{d_c^*(J)}. \tag{18}$$

The final weight associated to the provision can be expressed in the form

$$w_{cgi}^A = ps f \frac{1}{\pi_{cg}^A} \gamma_{cg} \frac{1}{\pi_{i|cg}^A}. \tag{19}$$

Table 1 shows the distribution of centers (total and respondents) and homeless persons interviewed by geographical area and type of center.

4.3.3 Link Reconstruction

Proper mapping of the individual links between units of the two populations is an important and delicate phase of the estimation procedure, because through this operation it is possible associate to each unit k selected in the sample s^B the weights of the provisions linked to it.

The links between the two populations were reconstructed, as mentioned above, on the basis of information collected in the questionnaire used for the interviews, that contained a diary section in which questions were asked about the centers frequented by the individual in the week prior to the day of the interview: for each day of the week, the interviewed person indicated the center, among those considered as the reference universe, in which he/she had lunch, dinner and/or slept.

In the survey the choice of the week as the observation period resulted to be a good decision. In fact, even though not all respondents were able to provide the retrospective information requested about their attendance of the centers in the week preceding the interview date, only 20 % of the cases presented a problem of identification of the link (link non-response).

In these cases it was necessary to treat the problem to avoid the risk of multiple counting of the same people.

The imputation of missing data in the diary was obtained through a probabilistic intra-record procedure, based on the proved fact that: (1) the behavior of homeless people in the use of services is regular; (2) the geographical characteristics and socio-demographic characteristics of homeless people with a partially completed diary, mostly related to the use of services, are not significantly different than the rest of the population.

The total mapping of links, consisting in the practical construction of the link matrix, allowed to assign to each individual the weights of the provisions w^A_{cgi} associated with the sample center-days cg in which he has been caught and to estimate the total number of links connecting the individual k to the population of provisions, $L^B_k(J)$.

4.4 Variance Estimation

The variance of estimator $\widehat{Y}^B(J) = \widehat{Z}^A(J)$ can be calculated on sample s^A using the equivalence between y^B and transformed variable z^A defined in (11).

Considering the sampling design adopted for the homeless people survey, variance estimation for estimator $\widehat{Y}^B(J)$ can be easily obtained through the expression derived for a two-stage sampling scheme with equal probabilities, both at first and second stage. This last condition is valid only approximately, as inclusion probabilities are nearly constant.

In order to describe the application context, it is necessary to introduce the following symbolic notation (for sake of brevity the reference period of the survey J is not indicated):

- N^A_{cg}: Number of provisions (secondary sample units) belonging to day g (primary sample units—opening days) of center c (stratum).
- n^A_{cg}: Number of selected provisions belonging to day g of center c.
- z^A_{cgi}: Value of the variable z^A on provision i belonging to day g of center c.
- $Z^A_{cg} = \sum_{i=1}^{N^A_{cg}} z^A_{cgi}$: Total of the variable z^A calculated in day g belonging to center c.

The estimator $\widehat{Z}^A$ defined in formula (13), expressed considering the two sampling stages as

$$\widehat{Z}^A = \sum_{c=1}^{C}\sum_{g=1}^{d_c}\sum_{i=1}^{n_{cg}^A} \frac{z_{cgi}^A}{\pi_{cgi}^A}, \tag{20}$$

can be rewritten—substituting in (20) the inclusion probabilities defined in Sect. 3.6—in the form

$$\widehat{Z}^A = \sum_{c=1}^{C}\frac{D_c}{d_c}\sum_{g=1}^{d_c}\frac{N_{cg}^A}{n_{cg}^A}\sum_{i=1}^{n_{cg}^A} z_{cgi}^A = \sum_{c=1}^{C}\frac{D_c}{d_c}\sum_{g=1}^{d_c}\widehat{Z}_{cg}^A = \sum_{c=1}^{C}\widehat{Z}_c^A, \tag{21}$$

in which:

$$\widehat{Z}_{cg}^A = \frac{N_{cg}^A}{n_{cg}^A}\sum_{i=1}^{n_{cg}^A} z_{cgi}^A \quad\text{and}\quad \widehat{Z}_c^A = \frac{D_c}{d_c}\sum_{g=1}^{d_c}\widehat{Z}_{cg}^A. \tag{22}$$

An estimator of the variance of $\widehat{Z}^A$ (Cicchitelli et al. 1992) is given by

$$\widehat{V}\left(\widehat{Z}^A\right) = \sum_{c=1}^{C}\widehat{V}_c\left(\widehat{Z}_c^A\right) = \sum_{c=1}^{C}D_c^2\frac{D_c - d_c}{D_c}\frac{1}{d_c}\frac{\sum_{g=1}^{d_c}\left(\widehat{Z}_{cg}^A - \sum_{g=1}^{d_c}\widehat{Z}_{cg}^A/d_c\right)}{d_c - 1}$$

$$+\sum_{c=1}^{C}\frac{D_c}{d_c}\sum_{g=1}^{d_c}\left(N_{cg}^A\right)^2\frac{N_{cg}^A - n_{cg}^A}{N_{cg}^A}\frac{1}{n_{cg}^A}\widehat{S}_{cg}^2, \tag{23}$$

where $\widehat{S}_{cg}^2$ is the estimate of second-stage variance of z^A in sample center-day cg

$$\widehat{S}_{cg}^2 = \frac{1}{n_{cg}^A - 1}\sum_{i=1}^{n_{cg}^A}\left(z_{cgi}^A - \sum_{i=1}^{n_{cg}^A} z_{cgi}^A/n_{cg}^A\right)^2. \tag{24}$$

From (23) it can be observed that the variance of estimator $\widehat{Z}^A$ (equivalent to $\widehat{Y}^B$) depends on first-stage variability (deriving from selection of sample days) and on variability of links collected on individuals associated to selected provisions. Actually, the second source of sampling variability is connected to the attendance of centers of homeless people.

5   Main Results of the Survey

During 30 days between November and December 2011 an estimated 47,648 homeless people used a canteen or night-time accommodation service at least once in the main 158 Italian municipalities (ISTAT 2013).

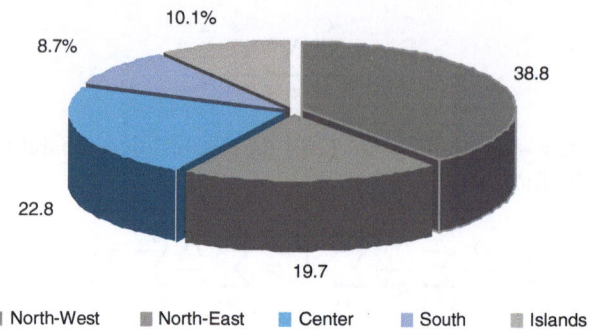

10.1%
8.7%
38.8
22.8
19.7

■ North-West   ■ North-East   ■ Center   ■ South   ■ Islands

Fig. 1 Distribution of homeless people in Italian geographical areas

Figure 1 shows the percentage distribution of homeless people in Italian geographical areas. As it was expected, most of the homeless people are concentrated in the North and in the Center where there is a strong presence of services provided to them.

The higher percentages observed in the North-west and in Central Italy essentially depend on the fact that Milano and Roma account for as much as 71 % of the samples surveyed. As many as 44 % of homeless people use services based in Roma (16.4 %) or Milano (27.5 %).

Most homeless people were male (86.9 %) and under the age of 45 (57.9 %), the majority were foreign citizens (59.4 %).

Table 2 also shows significant results that highlight some characteristics of homeless people and the differences that exist between Italians and foreign homeless people. The results refer only to homeless people capable of answering the interview, whose estimate amounts to 43,219 people. More than a quarter (28.3 %) are in employment particularly occasional or temporary work (24.5 %) and in low-qualified jobs. 71.7 % of homeless people did not work at all, more than half of homeless people (51.5 %) stated that they did not work because they couldn't find a job. Only 6.7 % had never had a job (a quarter of these were female, two-thirds were foreign citizens and under the age of 35).

The loss of employment was one of the most important factors in the gradual process of social exclusion that leads to "homelessness," along with separation from spouses and/or children and, to a more limited extent, health issues. As many as 61.9 % of homeless people had lost a stable employment position, 59.5 % had separated from their spouse and/or children and 16.2 % stated they were in bad or very bad health. Moreover, very few had not experienced any or only one of these events, confirming the fact that homelessness is caused by a combination of factors.

Table 2 Homeless people[a] by citizenship and other characteristics—Year 2011

	Foreign citizen	Italian	Total
Gender			
Male	87.6	86.2	87.0
Female	12.4	13.9	13.0
Age group			
18–34	46.5	10.4	31.8
35–44	27.7	22.0	25.3
45–54	17.4	30.3	22.7
55–64	7.0	26.5	14.9
>65	1.4	10.8	5.3
Employment			
Employed	27.8	29.2	28.3
Unemployed	72.2	70.8	71.7
– Unemployed never been employed	*7.7*	*5.4*	*6.7*
Type of event (not mutually exclusive)			
Disease	13.7	19.8	16.2
Separation from spouse and/or children	54.4	67.0	59.5
Loss of stable employment	55.9	70.6	61.9

[a]Percentage column composition, except for the last part. The results are referred to homeless people capable of answering the interview

6 Conclusions

The adopted approach represents an important innovation for the Italian official statistics because of two main reasons: for the first time the homeless population is surveyed at a national level on the whole Italian territory; and a new methodological instrument, such as indirect sampling, is experimented for a large-scale survey, obtaining very encouraging results.

The analysis of the outcome of the methodological approach has highlighted some features of the implementation which can be taken into account for improving the future realization of the survey on the homeless, which will be carried out in 2014. In particular, one relevant aspect is that for future applications it will be possible to exploit the information about the variability of links, associated to interviewed individuals, and of the flow of users, in order to plan a more efficient sampling design. These critical aspects are examples of what can be improved on the basis of past experience. Other limitations of the methodological approach, such as coverage issues, can be overcome only by conducting control investigations using different survey techniques.

References

Ardilly, P., Le Blanc, D.: Sampling and weighting a survey of homeless persons: a French example. Surv. Methodol. **27**(1), 109–118 (2001a)

Ardilly, P., Le Blanc, D.: Échantillonnage et pondération d'une enquête auprès de personnes sans domicile: un exemple français. Techniques d'enquête **27**(1), 117–127 (2001b)

Cicchitelli, G., Herzel, A., Montanari, G.E.: Il campionamento statistico. Il Mulino, Bologna (1992)

Dennis, M.L., Iachan, R.: A multiple frame approach to sampling the homeless and transient population. J. Off. Stat. **14**(5), 1–18 (1993)

Deville, J.C., Lavallée, P.: Indirect sampling: the foundation of the generalized weight share method. Surv. Methodol. **32**(2), 165–176 (2006)

Istat: Services to homeless people. http://www.istat.it/en/archive/45837 (2011a). 21 Nov 2011

Istat: I servizi alle persone senza dimora. http://www.istat.it/it/archivio/44096 (2011b). 3 novembre 2011

Istat: The homeless. http://www.istat.it/en/archive/92503 (2013). 10 Jun 2013

Kalton, G.: Practical methods for sampling rare and mobile populations. In: Proceedings of the Annual Meeting of the American Statistical Association, Atlanta, 5–9 August 2001

Lavallée, P.: Correcting for non-response in indirect sampling. In: Statistics Canada International Symposium Series – Proceedings (11-522-X) (2001)

Lavallée, P.: Indirect Sampling. Springer, New York (2007)

Marpsat, M., Razafindratsima, N.: Survey methods for hard-to-reach populations: introduction to the special issue. Methodol. Innov. Online **5**(2), 3–16 (2010)

Pannuzi, N., Masi, A., Siciliani, I.: Surveys of homeless persons – the case of Italy, counting the homeless. Peer Review, 12–13 November (2009)

Xiaojian, X., Lavallée, P.: Treatments for link nonresponse in indirect sampling. Surv. Methodol. **35**(2), 153–164 (2009)

Modelling Measurement Errors by Object-Oriented Bayesian Networks: An Application to 2008 SHIW

Daniela Marella and Paola Vicard

Abstract In this paper we propose to use the object-oriented Bayesian network (OOBN) architecture to model measurement errors. We then apply our model to the Italian survey on household income and wealth (SHIW) 2008. Attention is focused on errors caused by the respondents. The parameters of the error model are estimated using a validation sample. The network is used to stochastically impute micro data for households. In particular imputation is performed also using an auxiliary variable. Indices are calculated to evaluate the performance of the correction procedure and show that accounting for auxiliary information improves the results. Finally, potentialities and possible extensions of the Bayesian network approach both to the measurement error context and to official statistics problems in general are discussed.

Keywords Bayesian networks • Measurement error • Respondent error

1 Introduction

Measurement errors occur when the value provided by the respondents differs from the real but unknown value of a variable of interest in a survey. Such errors may arise at the data collection stage for different reasons. They may be due to the respondent, the interviewer, the wording of the questions in the questionnaire and the mode used to obtain the responses. The use of standard analysis in the presence of measurement

D. Marella (✉)
Dipartimento di Scienze della Formazione, Università Roma Tre, via del Castro Pretorio 20, 00185 Rome, Italy
e-mail: daniela.marella@uniroma3.it

P. Vicard
Dipartimento di Economia, Università Roma Tre, via Silvio D'Amico 77, 00145 Rome, Italy
e-mail: paola.vicard@uniroma3.it

F. Mecatti et al. (eds.), *Contributions to Sampling Statistics*, Contributions to Statistics, DOI 10.1007/978-3-319-05320-2_9,
© Springer International Publishing Switzerland 2014

errors usually gives misleading results leading to erroneous conclusions. Therefore it is important to have models for the measurement errors generating mechanism by which reconstructing individual true values. In this way, a predicted data set is produced on which standard inferential techniques can be applied. Basic contributions to the measurement errors models in survey sampling were given by Mahalanobis (1946), Hansen et al. (1961), Fuller (1987), Biemer (2009), and Lessler (1984).

This paper is mainly focused on modelling the error generating mechanism affecting ordinal categorical variables in order to predict corrected values. In particular, measurement errors in categorical variables due to the respondent are described by a mixed measurement model implemented in an object-oriented Bayesian network (OOBN), see Marella and Vicard (2013) for details. Such a network can then be used both to graphically and statistically model the error generating process and to stochastically impute micro data for units in order to reconstruct the unknown true values. More specifically, the evidence given by the observed value of the variable of interest is inserted and propagated throughout the network to estimate, for each unit, the probability distribution of the true value given the observed one. True values can be predicted by a random draw from such a distribution.

Our methodology is applied to data from the 2008 survey on household income and wealth (SHIW, for short). SHIW is conducted by Banca d'Italia every 2 years. Its main objective is to study the economic behaviors of Italian households. Interviews are considered valid if they have no missing items on the questions regarding income and wealth (Neri and Ranalli 2011). Therefore unit nonresponse and measurement errors are the principal sources of nonsampling errors that can severely bias the estimates. Here we deal with the measurement error. In fact, financial assets in SHIW are affected by misreporting of both the financial tools and the respective amounts with a prevalence of underreporting.

The paper is organized as follows. In Sect. 2 the mixed measurement model is described. Section 3 deals with the network implementing the overall measurement process. In Sect. 4 the network is used to impute micro data in 2008 SHIW data and the performance of the imputation procedure is evaluated. Finally, in Sect. 5 potentialities and possible extensions of our approach are discussed.

2   The Mixed Measurement Model

We focus on classification error in an ordered categorical variable due to the respondent providing a false answer. The respondent may either consciously or unconsciously provide incorrect answers, that may be due to the respondent willingness to misreport, to misunderstanding the question or not knowing the true answer.

Let X be an ordered categorical variable with K response categories whose population frequencies p_k, $k = 1, \ldots, K$, are assumed known from administrative archives or census data or estimated via *ad hoc* pilot survey. When a measurement error takes place the observed category for a given unit is different from the true category. Let $q_{i \to j}$ be the intercategory transition probability from the true category i to the observed category j, where $\sum_{j=1}^{K} q_{i \to j} = 1$.

In order to estimate the $K(K-1)$ probabilities $q_{i \to j}$, we could carry out an interview–reinterview study. An alternative way to proceed is to express the transition probabilities $q_{i \to j}$ by means of models characterized by a smaller number of parameters to be estimated. Here we use the scalar mutation model (see Vicard and Dawid 2004) in the measurement error context and call it *scalar measurement model*. This gives a simple but flexible description of the measurement error process, and it is given by

$$q_{i \to j} = \lambda s_{i \to j} \tag{1}$$

where $s_{i \to j}$, $j \neq i$ and λ are the unknown measurement parameters to be estimated. The parameter λ will be called *Error parameter*. Since $\lambda \sum_{j:j \neq i} s_{i \to j}$ is the probability of a measurement error when i is the true category, the overall measurement error rate μ can be expressed in terms of λ as follows:

$$\mu = \sum_{i} p_i \lambda \sum_{j:j \neq i} s_{i \to j}$$
$$= \beta \lambda \tag{2}$$

where

$$\beta = \sum_{i} \sum_{j \neq i} p_i s_{i \to j}. \tag{3}$$

Different measurement models can be identified by specifying the nonnegative quantities $s_{i \to j}$.

A realistic and plausible representation of the measurement error generating process is the *mixed measurement model* that has

$$s_{i \to j}^{\text{mix}} = (1 - h)s_{i \to j}^{\text{prop}} + h s_{i \to j}^{\text{MMT}} \tag{4}$$

given by a mixture of the one-T step model (MMT) and the proportional model (prop), where the mixture parameter h (for h between 0 and 1) reflects the relative importance of each model component. More specifically, the *proportional measurement model* is given by

$$s_{i \to j}^{\text{prop}} = p_j, i \neq j \tag{5}$$

that is, the larger the frequency p_j of category j, the more likely category i mutates into category j. Such a model reflects the assumption that whenever a measurement error occurs, the observed value is generated at random from the

population frequency distribution. This model is particularly suited for cases where
the respondent either unconsciously gives the false answer (not knowing the true
one) or decides to hide himself/herself in the population of interest answering
according to the frequency distribution.

The one-T step model MMT reflects the assumption that in real cases most
measurement errors occur in one of the T nearest neighbouring categories. For
instance, for $T = 2$ we obtain the *one-two step measurement model* given by

$$
s_{i \to j}^{MM2} = \begin{cases}
\alpha_1^- & \text{if } (i - j) = 1, \ i \neq 1 \\[2mm]
\alpha_1^+ & \text{if } (i - j) = -1, \ i \neq K \\[2mm]
\alpha_2^- & \text{if } (i - j) = 2, \ i \neq 1, 2 \\[2mm]
\alpha_2^+ & \text{if } (i - j) = -2, \quad i \neq K - 1, K \\[2mm]
\alpha_1 & \text{if } |i - j| = 1, \ i = 1 \quad \text{or} \quad i = K \\[2mm]
\alpha_2 \ \text{if } \ |i - j| = 2, \ i = 1, 2 \quad \text{or} \quad i = K - 1, K \\[2mm]
0 & \text{otherwise}
\end{cases}
\tag{6}
$$

where $\alpha_1 = \alpha_1^- + \alpha_1^+$, $\alpha_2 = \alpha_2^- + \alpha_2^+$ and $\alpha_1 + \alpha_2 = 1$. The model (6) implies
that the observed category can only be a neighbouring category or a category two
steps away. This model suitably represents those respondents that willing to give a
false but plausible answer, declare values not too far from the true ones. The one-
T step model is also useful when the observed variables are mainly affected by
either underreporting or overreporting. For instance, when income or financial assets
are investigated, there generally is a sensible prevalence of underreporting cases
with larger entity in the mismatch between the declared and the true values. These
asymmetric situations can be encoded in the one-T step model. In this case, it can
happen that the number of steps backward and forward may be different (the first
one being larger when income or wealth are analysed). For simplicity of notation
and to make the different entities of mismatch clear, we will denote the maximal
difference in the number of categories moving backward and forward from the true
category by T and T', respectively.

3   OOBN Implementing the Mixed Measurement Model

In order to automate and efficiently perform error detection and correction, the
mixed measurement model (4) has been implemented in a OOBN. For an account
on OOBNs, we refer to Koller and Pfeffer (1997).

Fig. 1 Top-level network for the measurement error model for the respondent

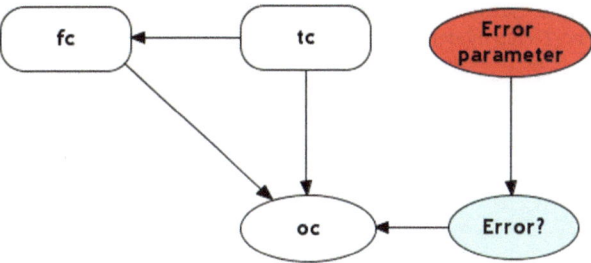

Fig. 2 Subnetwork for the false category

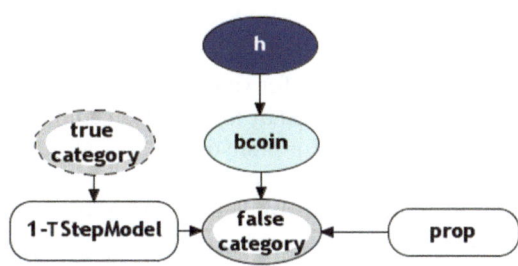

Figure 1 shows the main network (top-level) representing the overall measurement error process. In what follows, instance and regular nodes are indicated in teletype face, while boldface is used for network classes.

In Fig. 1 the observed category is represented by the standard node oc; the true category tc is represented by an instance of network class **tc** associated with the probability distribution of the variable of interest. The fact that the respondent may be a liar is represented by the random node **Error?** associated with a Bernoulli distribution of parameter λ, where from (2)

$$\lambda = \mu/\beta \quad \text{with} \quad \beta = (1-h)\left(1 - \sum_i p_i^2\right) + h$$

i.e. λ is a linear transformation of the overall measurement error rate μ. More specifically, if the respondent is *honest*, coded as Error? = 0, then the observed category coincides with the original one. If the respondent is a *liar*, coded as Error? = 1, then the observed category is different from the original one and it is generated according to the mixed measurement model. In particular, the false category (node fc) is represented by an instance of network class **model** that is itself a Bayesian network (shown in Fig. 2).

The network in Fig. 2 encodes the mixed measurement model (4). Thus, the false category is chosen as either a category (prop node) generated by the proportional model (5) or a category (1-TStepModel node) generated by the one-T step model (6) according to a biased coin toss. The biased coin toss is represented by node bcoin that is associated with a Bernoulli variable of parameter h (node h), i.e. the mixture parameter in (4).

Fig. 3 Subnetwork for the
one-T step model

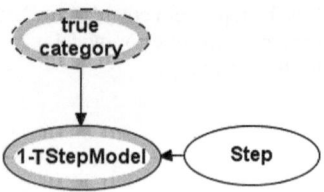

Going deeper into the hierarchical structure of the OOBN, the one-T step measurement model is encoded in a separate subnetwork (Fig. 3). Specifically, node `1-TStepModel` in Fig. 2 is an instance of class **one-T-step** represented in Fig. 3. The node `Step` is associated with a probability distribution with as many states as the possible steps forward and backward with respect to the true category, and probabilities estimated by a validation sample (if available) or through an interview–reinterview study. For more details, see Marella and Vicard (2013).

It is important to stress that the modularity property of BNs (Cowell et al. 2007) and the hierarchical structure of OOBNs allow to model the mixed measurement model (4) in a separate subnetwork improving the readability of the overall network. Moreover, thanks to OOBN hierarchical architecture, on one hand the final user may be interested in the overall error generating mechanism, without focusing on the statistical model specification, and therefore he/she only needs the top-level network in Fig. 1. On the other hand, a deeper analysis of the model details can be performed simply "looking inside" the network class **model** for the measurement model in Fig. 2 and eventually (if necessary) "magnifying" the node `1-TStepModel` in order to check the one-T step model parameters specification.

4 An Application to 2008 SHIW

The performance of our OOBN model, together with its sensitivity to the parameter values, has been evaluated by means of simulated data (Marella and Vicard 2013). Here we apply our model to the 2008 Banca d'Italia SHIW.

4.1 The Data

SHIW is a survey conducted by Banca d'Italia every 2 years. It collects detailed information on household characteristics, and its members, such as income (for individual and households), savings and consumptions, wealth in terms of financial assets and liabilities. SHIW is a two stage survey, with municipalities and households as primary and secondary sampling units, respectively. The municipalities are stratified by administrative region aggregations (NUTS 1 level, i.e. Northwestern, Northeastern, Central, Southern and Insular Italy) and population size (less than

20,000 inhabitants; between 20,000 and 40,000; 40,000 or more). Within each stratum, municipalities are selected to include all those with a population of 40,000 inhabitants or more and those with panel households (self-representing municipalities), while smaller municipalities are selected using probability proportional to size sampling (without replacement). Individual households are then randomly selected from administrative registers.

A validation sample has been used to estimate the measurement model parameters. Data in the validation sample have been collected through an independent experiment survey carried out by Banca d'Italia and a major Italian bank group on a sample of customers of the latter. The survey was carried out in 2003 on a sample of 1,681 households where at least one member was a customer of the bank group. Survey data have then been matched with the bank customers database containing the amount of the assets actually held by the individuals selected in the sample. The resulting dataset will be referred to as our validation sample, see Neri and Ranalli (2011) for details.

4.2 Results of Micro Data Imputation

In our analysis we focus on bonds amount (specifically government and private bonds). Since the mixed measurement model describes measurement error in a categorical variable, the true and the observed amount of bond in the validation sample have been discretized in ten categories. The discretization has been performed using *Chi2* discretization algorithm, see Liu and Setiono (1995). In the following we treat the discretized distribution as the true distribution.

A two step procedure has been followed:

1. First of all, the measurement error rate μ and the one-T step model parameters $(\alpha_t^+, \alpha_{t'}^-)$, for $t = 1, \dots, T$ and $t' = 1, \dots, T'$, have been estimated using the true and the observed data in the validation sample.

   The numbers of steps forward and backward have been established on the basis of the length of the mismatches (in terms of number of classes) between the declared and the true bond amount class in the validation sample.

   It resulted that the maximal underreporting length (T) is 7, while the maximal overreporting length (T') is 3.
2. Secondly, the estimated model suitably implemented in a BN is used to impute the discretized amount of bond in SHIW 2008.

As far as the first point is concerned, the misreport probability μ has been estimated as the proportion of mismatches between the observed and the true value of amount of bond in the validation sample. Analogously, the parameter α_t^+ and $\alpha_{t'}^-$ (for $t = 1, \dots, T$ and $t' = 1, \dots, T'$) representing respectively the probability that the difference between the observed and the true category is positive and equal to t or negative and equal to t', has been estimated by the proportion of mismatches between observed and true value of amount of bond of each considered length.

Table 1 Estimates of measurement model parameters

μ	α_7^-	α_6^-	α_5^-	α_4^-	α_3^-	α_2^-	α_1^-	α_1^+	α_2^+	α_3^+
0.54	0.07	0.07	0.08	0.15	0.12	0.13	0.16	0.11	0.06	0.05

Finally, the value of the mixture parameter $h = 0.9$ has been determined through a sensitivity analysis. The estimates of all the measurement model parameters are reported in Table 1.

The basic assumption in the imputation phase is that the underreporting behavior observed in the validation sample remained unchanged in 2008.

In order to evaluate the performance of the imputation procedure, the following evaluation criteria have been used:

1. The Kullback–Leibler distance between the true distribution and the imputed one and between the true distribution and the observed one, denoted by KL^{TI} and KL^{TO}, respectively.
2. The percentage of correct imputations defined as

$$\psi = \frac{1}{n^*} \sum_{i \in S^*} I_{x_i}(x_i^*) * 100 \qquad (7)$$

where S^* is the subsample composed of units affected by measurement errors in the validation sample and I_{x_i} is the indicator function assuming the values 1 if the true value for unit i denoted by x_i is equal to the corresponding imputed value x_i^* and 0 otherwise.

The imputation procedure reveals a good performance when applied to categorical variables. The distance between the true and the imputed distribution $KL^{TI} = 0.13$ is less than the distance between the true and the observed distribution $KL^{TO} = 0.47$. Furthermore, the proposed imputation method is able to correctly reconstruct 11 % of data affected by measurement error in the validation sample.

In Fig. 4 the distribution of true, observed and imputed values are reported. First of all, the distribution of observed values is characterized by positive skewness due to underreporting in the amount of bond. Furthermore, the imputation procedure partially corrects the bias in the distribution of observed values due to the presence of measurement errors.

4.2.1 Using Auxiliary Information

An alternative approach consists in estimating the misreport probability by proportions of mismatches between observed and true value of amount of bond in the validation sample within groups. The groups are mutually exclusive and exhaustive and are defined with the aid of auxiliary information. Formally, the validation sample is partitioned into $H(s)$ groups s_h $(h = 1, \ldots, H(s))$ on the basis of

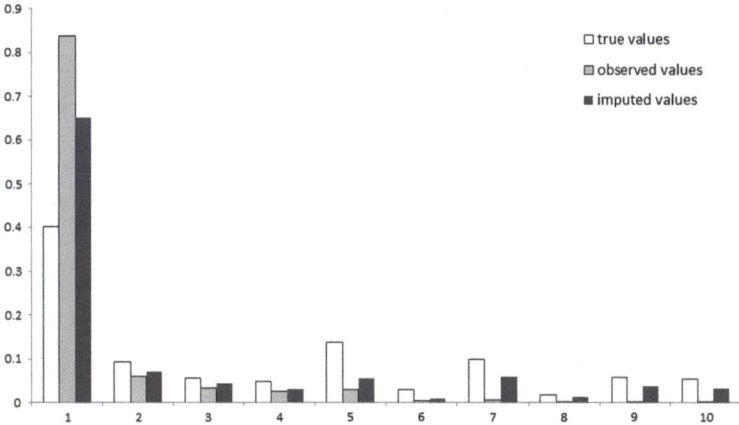

Fig. 4 Distribution of true, observed and imputed values

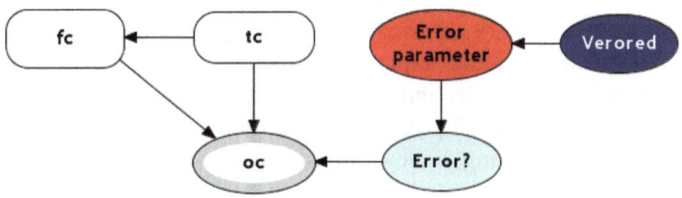

Fig. 5 Top-level network for the measurement error model for the respondent when auxiliary information on variable *Verored* is considered

auxiliary information such that all elements within the same group are assumed to have the same misreport probability μ_h. The sets $(s_1, \ldots, s_h, \ldots, s_{H(s)})$ are the misreport homogeneity groups. Let l_h be the misreporting subset of the group s_h, and let us denote by n_h and m_h the sizes of s_h and l_h, respectively; then the individual misreport probability for unit i belonging to the group s_h (μ_{ih}) is

$$\mu_{ih} = \mu_h = \frac{m_h}{n_h} \tag{8}$$

that is the probability of misreporting of any element i in group h is equal to the group misreport rate.

In our application we consider the information provided by the categorical variable *Verored*. The variable, characterized by ten classes of response, is compiled by the interviewer that must express a judgment on the reliability of the information provided by the interviewee about income and wealth.

It is important to notice that the inclusion of such auxiliary information in the OOBN model is straightforward. The resulting network is shown in Fig. 5. In fact, thanks to modularity, it is sufficient to add the node Verored representing the classification of the validation sample units in the ten misreport homogeneity groups.

Table 2 Estimates of misreport probability within groups defined by variable *Verored*

μ_1	μ_2	μ_3	μ_4	μ_5	μ_6	μ_7	μ_8	μ_9	μ_{10}
0.74	0.49	0.60	0.51	0.60	0.58	0.49	0.55	0.52	0.48

Table 3 Estimates of misreport probabilities within groups defined by variable *Education*

	Degree	Diploma	Other
True/observed	0.68	0.60	0.56
True/imputed	0.61	0.53	0.50

Then the conditional probability table associated with node Error parameter simply chooses out of the possible ten error parameters the one corresponding to each specific *Verored* group.

The estimates of the misreport probabilities within each group are reported in Table 2.

The distance between the true and the imputed distribution decreases from 0.13 to 0.10 when auxiliary information represented by the variable *Verored* is considered. The results of the imputation procedure could be further improved defining household profiles as unique combinations of covariates values when adequate sample sizes for each profile are available. Also in this case the network can be modified just adding a new module relative to the set of considered covariates.

In our application the information provided by the categorical variable *Verored* has been considered. Such a variable has been chosen on the basis of its ability to correct the data, i.e. the larger is the reduction of the Kullback–Leibler distance between the distribution of true and imputed values, the larger is the data correction ability of the auxiliary variable.

It could be interesting to check if the data correction due to *Verored* preserves the correlation between the variable of interest and the available socio-demographic auxiliary information. As an example, we considered the variable *Education*. Broadly speaking, it is plausible to expect that measurement error is more frequent in people with high education. This is confirmed in our application as shown in Table 3 where the misreport probability has been estimated within education groups (*degree*, *diploma*, *other*) as the mismatches between the true and the observed values. In Table 3 the mismatches between the true and the imputed values computed for the different education groups are also reported. It results that the correction process:

1. Decreases the misreport probabilities within each education group exploiting the correlation between *Education* and *Verored*.
2. Preserves the correlation structure between the misreport propensity and the education level.

5   Conclusions

Here we have shown how BNs and their extension OOBNs can be used to tackle the problem of measurement error. An OOBN has been built to model the measurement error in one discrete variable. Model parameters have been estimated and micro data imputation has been performed. The results are encouraging showing a remarkable decrease in Kullback–Leibler distance from the true distribution when imputed data are used instead of observed data. Therefore we believe that BNs are an important and promising tool to deal with measurement error in sample surveys.

In this application we have also shown that results are improved when auxiliary information is taken into account. Therefore further research needs to be primarily focused on two extensions:

- Estimating the measurement error parameter using all the available auxiliary variables. In this example we have used only variable *Verored* but better results could be obtained if all the variables carrying information about the propensity to lie or to commit an error would be considered. To this aim it is necessary to learn a BN, i.e. derive the association structure, of the variables in the validation sample.
- Studying whether it is possible to improve micro data imputation results taking into account those variables directly influencing the one to be corrected.

BNs have been already successfully applied to official statistics problems such as imputation (Thibaudeau and Winkler 2002; Di Zio et al. 2004, 2005), contingency table estimation (Ballin et al. 2010) and paradata analysis (Ballin et al. 2006). However, there are still some limitations that may complicate their application. The two main problems regarding also the measurement error model context are:

1. The use of hybrid BNs where continuous and discrete variables are considered.
2. The necessity to take into account the complexity of sampling design when BNs are applied to sample surveys.

With regard to the first point, exact probabilistic inference is possible in networks with Conditional Gaussian (CG) density functions (Cowell et al. 2007). However, in SHIW data we have dependencies substantially different from those encompassed by the CG model, i.e. the continuous variables are characterized by skew to the right distributions, and then the Gaussian assumption is not applicable.

The most commonly used approach in practice is to discretize all continuous variables. An alternative approach could be to use mixture of BNs (see Shenoy 2006).

As far as the second point is concerned, the sampling design complexity must be taken into account by an appropriate use of sampling weights both in the estimation of measurement model parameters and in the structural learning algorithm.

References

Ballin, M., Scanu, M., Vicard, P.: Paradata and Bayesian networks: A tool for monitoring and troubleshooting the data production process. In: Working Paper no 66 - 2006. Dipartimento di Economia Università Roma Tre (2006). Available via DIALOG. http://host.uniroma3.it/dipartimenti/economia/pdf/wp66.pdf

Ballin, M., Scanu, M., Vicard, P.: Estimation of contingency tables in complex survey sampling using probabilistic expert systems. J. Stat. Plan. Inference **140**, 1501–1512 (2010)

Biemer, P.P.: Measurement errors in sample survey. In: Pfeffermann, D., Rao, C.R. (eds.) Handbook of Statistics, vol. 29A. Sample Surveys: Design, Methods and Applications. North-Holland, Amsterdam (2009)

Cowell, R.G., Dawid, A.P., Lauritzen, S.L., Spiegelhalter, D.J.: Probabilistic Networks and Expert Systems: Exact Computational Methods for Bayesian Networks, 1st edn. Springer Publishing Company, Incorporated (2007)

Di Zio, M., Scanu, M., Coppola, L., Luzi, O., Ponti, A.: Bayesian networks for imputation. J. R. Stat. Soc. A **167**(2), 309–322 (2004)

Di Zio, M., Sacco, G., Scanu, M., Vicard, P.: Multivariate techniques for imputation based on Bayesian networks. Neural Netw. World **4**, 303–309 (2005)

Fuller, W.A.: Measurement Error Models. Wiley, New York (1987)

Hansen, M.H., Hurwitz, W.N., Bershad, M.A.: Measurement errors in censuses and surveys. Bull. Int. Stat. Inst. **38**, 359–374 (1961)

Koller, D., Pfeffer, A.: Object-oriented Bayesian networks. In: Proceedings of the Thirteenth Annual Conference on Uncertainty in Artificial Intelligence, pp. 302–313 (1997)

Lessler, J.T.: Measurement errors in surveys. In: Turner, C.F., Martin, E. (eds.) Surveying Subjective Phenomena, vol. 2, pp. 405–440. Russel Sage Foundation, New York (1984)

Liu, H., Setiono, R.: Chi2: Feature selection and discretization of numeric attributes. In: Proceedings of The Seventh International Conference with Artificial Intelligence (1995)

Mahalanobis, P.C.: Recent experiments in statistical sampling in the Indian Statisticla Institute. J. R. Stat. Soc. **109**, 325–370 (1946)

Marella, D., Vicard, P.: Object-oriented Bayesian network for modeling the repsondent measurement error. Commun. Stat. Theory Methods **42:19**, 3463–3477 (2013)

Neri, A., Ranalli, M.G.: To misreport or not to report? The case of the Italian survey on household income and wealth. Stat. Transit. **12**, 281–300 (2011)

Shenoy, P.P.: Inference in hybrid Bayesian networks using mixtures of Gaussians. In: Proocedings of the 22nd Conference in Uncertainty in Artificial Intelligence (2006)

Thibaudeau, Y., Winkler, W.E.: Bayesian networks representations, generalized imputation, and synthetic micro-data satisfying analytic constraints. In: Research Report RRS2002/9 - 2002, U.S. Bureau of the Census (2002). Available via DIALOG. www.census.gov/srd/papers/pdf/rrs2002-09.pdf.

Vicard, P., Dawid, A.P.: A statistical treatment of biases affecting the estimation of mutation rates. Mutat. Res. **547**, 19–33 (2004)

Sampling Theory and Geostatistics: A Way of Reconciliation

Giorgio E. Montanari and Giuseppe Cicchitelli

Abstract In a previous paper, the authors addressed the problem of the estimation of the mean of a spatial population in a design-based context and proposed a model-assisted estimator based on penalized regression splines. This paper goes a step further exploring the performance of the leave-one-out cross-validation criterion for selecting the amount of smoothing in the regression spline assisting model and for estimating the variance of the proposed estimator. The attention is focused on continuous spatial populations. A simulation study provides an empirical efficiency comparison between this estimator and the kriging predictor by means of Monte Carlo experiments.

Keywords Cross-validation • Kriging • Model-assisted estimator • Non-parametric estimation • Penalized splines

1 Introduction

In environmental studies, dealing with soil characterization, monitoring of natural resources or estimation of pollution concentration, often it is of interest to estimate the mean of a response variable defined continuously over a geographical region.

Let $A \subset R^2$ be a planar domain of interest and let $y(\mathbf{x})$ be the response variable defined for all locations $\mathbf{x} = (x_1, x_2)$ in A, where x_1 and x_2 are the geographical coordinates of the location. A common problem in this context is to estimate the mean of the response variable, using values of $y(\mathbf{x})$ observed in an ordered sample of locations $s = (\mathbf{x}_1, \ldots, \mathbf{x}_n)$. We will focus on the case of a spatial continuous population and in such a case the population mean is given by

G.E. Montanari (✉) • G. Cicchitelli
University of Perugia, Perugia, Italy
e-mail: giorgio.montanari@unipg.it

F. Mecatti et al. (eds.), *Contributions to Sampling Statistics*, Contributions to Statistics,
DOI 10.1007/978-3-319-05320-2__10,
© Springer International Publishing Switzerland 2014

$$\bar{Y} = \frac{1}{|A|} \int_A y(\mathbf{x}) d\mathbf{x},$$

where $|A| = \int_A d\mathbf{x}$ is the area of domain A.

Classical sampling theory deals with finite populations. But sampling theory and continuous populations came into contact in the statistical analysis of spatial data. In fact, spatial points can be labelled and identified by their geographical coordinates, and this makes it possible to design sampling strategies to draw samples and to do inference on the population parameters. In this regard, pioneering contributions are those of McArthur (1987) and Cordy (1993). In the first paper, various sampling strategies for estimating the average level of a response surface representing a pollutant are compared. The second paper provides the theory for Horvitz–Thompson estimation of means or totals of continuous populations.

The problem of estimating the mean or the total of a continuous spatial population has been addressed, among others, by Brus (2000), Stevens and Jensen (2007) and Barabesi et al. (2012). These papers pursue the improvement of efficiency of the sampling strategy using, at the design stage, the auxiliary information provided by the spatial coordinates of location points and spatially balanced samples come out as optimal. In standard survey methodology, a sample is "balanced" with respect to a given auxiliary variable when the estimated mean of this variable equals the population mean. Higher order balancing can also be achieved. For example, we can require that the first two estimated moments (i.e., mean and variance) of the auxiliary variable match the population corresponding moments. This technique relies on the idea that when the auxiliary variables are correlated with the response variable, balancing over the auxiliary variables should produce an approximate balance over the unknown response, and a more precise estimate of the response population mean is expected.

For spatial populations, the most important auxiliary variable is the spatial location of points. Then, if we believe that the response variable has a spatial structure in the study domain, a balanced sample is one that is evenly and regularly spread out across the domain as much as possible (Stevens and Jensen 2007, p. 517).

There are various techniques to obtain spatially balanced samples. One of the most frequently used methods is spatial stratification by overimposing on the domain regular polygons or other tessellation criterions; we refer, in particular, to the random tessellation stratified (RTS) design and to the generalized random tessellation stratified (GRTS) design described in Stevens (1997) and in Stevens and Olsen (2004). These designs, coupled with the Horvitz–Thompson estimator of the mean, yield very efficient strategies compared with the standard benchmark represented by the uniform random sampling and the sample mean.

An alternative approach consists in using the auxiliary information provided by the locations of sample units at the estimation stage, looking for more efficient estimator than the Horvitz–Thompson one. In this regard, Cicchitelli and Montanari (2012), using a penalized spline regression model to capture the spatial pattern in the data, propose a design-consistent and approximately unbiased estimator of the population mean, within a model-assisted framework.

In this paper, we further develop the approach of Cicchitelli and Montanari focusing on the choice of the smoothing parameter for the penalized spline regression estimator. In particular, we report and discuss results from a wide simulation study carried out to explore this issue.

2 The Penalized Spline Regression Model-Assisted Estimator of the Population Mean

Let $\kappa_1, \ldots, \kappa_K$ be K locations in A, called knots, and let

$$\mathbf{z}(\mathbf{x}) = [z_1(\mathbf{x}), \ldots, z_K(\mathbf{x})] = \left[\breve{z}_1(\mathbf{x}), \ldots, \breve{z}_K(\mathbf{x}) \right] \mathbf{\Omega}^{-1/2}$$

be a row vector of covariates, where $\breve{z}_l(\mathbf{x}) = (||\mathbf{x} - \kappa_l||)^2 \log(||\mathbf{x} - \kappa_l||)$, with $l = 1, \ldots, K$, and $\mathbf{\Omega} = \{(||\kappa_l - \kappa_k||)^2 \log(||\kappa_l - \kappa_k||)\}_{k,l=1,\ldots,K}$.

Assume that the response variable $y(\mathbf{x})$ at location $\mathbf{x} \in A$ is a realization of the super-population model according to which

$$y(\mathbf{x}) = \beta_0 + \beta_1 x_1 + \beta_2 x_2 + \mathbf{z}(\mathbf{x})\mathbf{u} + \varepsilon(\mathbf{x}), \tag{1}$$

where $\beta_0, \beta_1, \beta_2$ and $\mathbf{u}' = [u_1, \ldots, u_K]$ are unknown regression parameters and $\varepsilon(\mathbf{x})$ is a white noise with variance σ_ε^2.

Assuming that the sample of locations s is randomly drawn from A with a fixed size n sampling design that assigns inclusion density $\pi(\mathbf{x})$ to location $\mathbf{x}$ (see Cordy 1993), Cicchitelli and Montanari (2012) introduce the design-consistent and approximately unbiased *p-spline regression estimator* (PSRE) of the population mean, which is given by

$$\hat{\bar{Y}}_{\text{spl}} = \hat{\beta}_0 + \hat{\beta}_1 \bar{x}_1 + \hat{\beta}_2 \bar{x}_2 + \bar{\mathbf{z}}(\mathbf{x})\hat{\mathbf{u}} + \frac{1}{|A|} \sum_{i=1}^{n} \frac{\hat{e}(\mathbf{x}_i)}{\pi(\mathbf{x}_i)}, \tag{2}$$

where $\bar{x}_1, \bar{x}_2$ and $\bar{\mathbf{z}}(\mathbf{x})$ are the population mean in A of x_1, x_2 and $\mathbf{z}(\mathbf{x})$; $\hat{\beta}_0, \hat{\beta}_1, \hat{\beta}_2$ and $\hat{\mathbf{u}}$ are the design-based estimates of the census parameters $\beta_{0P}, \beta_{1P}, \beta_{2P}, \mathbf{u}_P$ that would be obtained by fitting model (1) to the whole population by means of the penalized least square criterion which, in the continuous case, consists in minimizing

$$\int_A [y(\mathbf{x}) - \beta_0 - \beta_1 x_1 - \beta_2 x_2 - \mathbf{z}(\mathbf{x})\mathbf{u}]^2 d\mathbf{x} + \lambda \mathbf{u}'\mathbf{u},$$

for some fixed value of the penalizing parameter $\lambda \geq 0$; $\hat{e}(\mathbf{x}_i) = y(\mathbf{x}_i) - \hat{y}(\mathbf{x}_i)$ is the residual in $\mathbf{x}_i$, i.e. the difference between the observed value $y(\mathbf{x}_i)$ and the fitted value obtained from the function

$$\hat{y}(\mathbf{x}) = \hat{\beta}_0 + \hat{\beta}_1 x_1 + \hat{\beta}_2 x_2 + \mathbf{z}(\mathbf{x})\hat{\mathbf{u}}, \quad \mathbf{x} \in A. \tag{3}$$

The estimated census parameters in (2) are given by

$$\begin{bmatrix} \hat{\boldsymbol{\beta}} \\ \hat{\mathbf{u}} \end{bmatrix} = \left[\begin{bmatrix} \mathbf{X}_s' \mathbf{\Pi}_s^{-1} \mathbf{X}_s & \mathbf{X}_s' \mathbf{\Pi}_s^{-1} \mathbf{Z}_s \\ \mathbf{Z}_s' \mathbf{\Pi}_s^{-1} \mathbf{X}_s & \mathbf{Z}_s' \mathbf{\Pi}_s^{-1} \mathbf{Z}_s \end{bmatrix} + \lambda \mathbf{D} \right]^{-1} \begin{bmatrix} \mathbf{X}_s' \mathbf{\Pi}_s^{-1} \\ \mathbf{Z}_s' \mathbf{\Pi}_s^{-1} \end{bmatrix} \mathbf{y}_s, \tag{4}$$

where $\hat{\boldsymbol{\beta}}' = (\hat{\beta}_0, \hat{\beta}_1, \hat{\beta}_2)$; $\mathbf{X}_s$ is the matrix of order $n \times 3$ having as i-th row $(1, x_{i1}, x_{i2})$, for $i = 1, \ldots, n$; $\mathbf{Z}_s$ is the matrix of order $n \times K$ whose i-th row is given by $\mathbf{z}(\mathbf{x}_i)$; $\mathbf{\Pi}_s = \text{diag}\{\pi(\mathbf{x}_1), \ldots, \pi(\mathbf{x}_n)\}$ is the diagonal matrix having as elements the inclusion density values of sample locations; $\mathbf{D} = \text{blockdiag}\{\mathbf{0}_{3 \times 3}, \mathbf{I}_K\}$, where $\mathbf{I}_K$ is the identity matrix of order K and $\mathbf{0}_{3 \times 3}$ is a matrix of zeros of size indicated in the subindex; $\mathbf{y}_s = [y(\mathbf{x}_1), \ldots, y(\mathbf{x}_n)]'$ is the vector of the response sample values.

The approximate unbiasedness and design consistency of (2) have been proved assuming an asymptotic framework in which $n \to \infty$ and the ratio r/n goes to zero, being r the number of degrees of freedom used by the fitted model.

Quantity $\hat{\beta}_0 + \hat{\beta}_1 \bar{x}_1 + \hat{\beta}_2 \bar{x}_2 + \bar{\mathbf{z}}(\mathbf{x})\hat{\mathbf{u}}$ on the right-hand side of (2) is the population mean in A of $\hat{y}(\mathbf{x})$, which provides a design-based estimate of the response predicted value at location $\mathbf{x}$ that would be obtained with the census parameters. It is worthwhile noting that function (3) can be also used for mapping purposes. Furthermore, $|A|^{-1} \sum_1^n \hat{e}(\mathbf{x}_i)/\pi(\mathbf{x}_i)$ on the right-hand side of (2) is the Horvitz–Thompson estimator of the population mean of residuals. It is easy to recognize that (2) is a generalized regression (GREG) estimator for the continuous case. However, the weighted sum of the residuals can be shown to be equal to zero since model (1) contains the intercept and the variance of the error term $\varepsilon(\mathbf{x})$ is assumed to be constant (see Särndal et al. 1992, p. 231). So, the PSRE reduces to

$$\hat{\bar{Y}}_{spl} = \hat{\beta}_0 + \hat{\beta}_1 \bar{x}_1 + \hat{\beta}_2 \bar{x}_2 + \bar{\mathbf{z}}(\mathbf{x})\hat{\mathbf{u}}. \tag{5}$$

The covariates $\mathbf{z}(\mathbf{x})$ are radial basis functions for spline regression and this choice corresponds to the low rank *thin plate spline* family of smoothers (Ruppert et al. 2003). Note that function $(\|\mathbf{x} - \boldsymbol{\kappa}_l\|)^2 \log(\|\mathbf{x} - \boldsymbol{\kappa}_l\|)$ is a way to describe location $\mathbf{x}$ with respect to the knot $\boldsymbol{\kappa}_l$, and the vector $\mathbf{z}(\mathbf{x})$ is a sort of standardized measures that characterize the geographic position of $\mathbf{x}$ with respect to the K knots. In this way, the spatial information is conveyed into the spline basis, which depends on the spatial configuration of data and knots.

The relation between the penalty λ and the amount of smoothing can be expressed by the number of degrees of freedom used by the fitted model. Since the fitted values are given by $\mathbf{S}_{\lambda s} \mathbf{y}_s$, where

$$\mathbf{S}_{\lambda s} = [\mathbf{X}_s, \mathbf{Z}_s]\left[\begin{bmatrix} \mathbf{X}'_s\mathbf{\Pi}_s^{-1}\mathbf{X}_s & \mathbf{X}'_s\mathbf{\Pi}_s^{-1}\mathbf{Z}_s \\ \mathbf{Z}'_s\mathbf{\Pi}_s^{-1}\mathbf{X}_s & \mathbf{Z}'_s\mathbf{\Pi}_s^{-1}\mathbf{Z}_s \end{bmatrix} + \lambda\mathbf{D}\right]^{-1}\begin{bmatrix} \mathbf{X}'_s\mathbf{\Pi}_s^{-1} \\ \mathbf{Z}'_s\mathbf{\Pi}_s^{-1} \end{bmatrix}, \qquad (6)$$

the number of degrees of freedom, r, is given by the trace of the smoothing matrix, which can be also written as

$$r = \text{trace}\left\{\left[\begin{bmatrix} \mathbf{X}'_s\mathbf{\Pi}_s^{-1}\mathbf{X}_s & \mathbf{X}'_s\mathbf{\Pi}_s^{-1}\mathbf{Z}_s \\ \mathbf{Z}'_s\mathbf{\Pi}_s^{-1}\mathbf{X}_s & \mathbf{Z}'_s\mathbf{\Pi}_s^{-1}\mathbf{Z}_s \end{bmatrix} + \lambda\mathbf{D}\right]^{-1}\begin{bmatrix} \mathbf{X}'_s\mathbf{\Pi}_s^{-1}\mathbf{X}_s & \mathbf{X}'_s\mathbf{\Pi}_s^{-1}\mathbf{Z}_s \\ \mathbf{Z}'_s\mathbf{\Pi}_s^{-1}\mathbf{X}_s & \mathbf{Z}'_s\mathbf{\Pi}_s^{-1}\mathbf{Z}_s \end{bmatrix}\right\}.$$

It is easily seen that $3 \le r \le K+3$. The penalty parameter λ has the far most important impact on r: in fact, when $\lambda = 0$, then $r = K+3$, while r attains the minimum 3 as $\lambda \to \infty$. The value of r is sometimes called *equivalent number of parameters* of the smoother.

2.1 Variance Estimation and Penalizing Parameter

Our previous discussion has been embedded within the model-assisted approach to design-based inference. Estimator $\hat{Y}_{\text{spl}}$ is a GREG estimator and the standard design-consistent variance estimator for fixed sample size is given by

$$\hat{V}\left(\hat{Y}_{\text{spl}}\right) = \frac{1}{|A|^2}\sum_{i=1}^{n}\sum_{j>i}^{n}\frac{\pi(\mathbf{x}_i)\pi(\mathbf{x}_j) - \pi(\mathbf{x}_i, \mathbf{x}_j)}{\pi(\mathbf{x}_i, \mathbf{x}_j)}\left(\frac{\hat{e}(\mathbf{x}_i)}{\pi(\mathbf{x}_i)} - \frac{\hat{e}(\mathbf{x}_j)}{\pi(\mathbf{x}_j)}\right)^2, \qquad (7)$$

where $\pi(\mathbf{x}_k, \mathbf{x}_l)$ is the second-order inclusion density (Cordy 1993). This estimator would be design unbiased if the census parameters $\beta_{0P}, \beta_{1P}, \beta_{2P}$ and $\mathbf{u}_P$ were known. But when they are estimated, variance estimator (7) is downward biased since it does not include the variance component due to the estimation of the census parameters by means of $\hat{\beta}_0, \hat{\beta}_1, \hat{\beta}_2, \hat{\mathbf{u}}$. The bias may be substantial when the ratio between the sample size n and the number of degrees of freedom r is small. In fact, high values of r compared to n might cause an overfitting of sample data, and, as a consequence, might produce sample residuals much smaller than those for unsampled locations. In the simulation study in Cicchitelli and Montanari (2012), the 95 % confidence interval built with $\hat{Y}_{\text{spl}}$ and the variance estimator (7) has an effective coverage below 92 % as soon as the ratio n/r goes under 6. These results put a limit on the choice of the smoothing parameter λ and, therefore, on the number of degrees of freedom of the smoother, as long as estimator (7) is employed for estimating the variance.

The problem of optimal selection of the smoothing parameter has not been addressed by Cicchitelli and Montanari (2012). In this regard, Opsomer and Miller (2005) propose a *design-based* cross-validation (CV) criterion for selecting the bandwidth of their local polynomial regression estimator. Mimicking their approach, our proposal is to replace in (7) each $\hat{e}(\mathbf{x}_i)$ with the "leave-one-out" residual $\hat{e}_{(-i)}(\mathbf{x}_i)$, where the predicted response in $\mathbf{x}_i$ is computed using the regression function (3) estimated after having removed location $\mathbf{x}_i$ from the sample $s = (\mathbf{x}_1, \ldots, \mathbf{x}_n)$. Replacing $\hat{e}(\mathbf{x}_i)$ with $\hat{e}_{(-i)}(\mathbf{x}_i)$ in (7), we get the design-based CV criterion for choosing the smoothing parameter λ of the PSRE:

$$\hat{V}_{\mathrm{CV}}\left(\hat{\bar{Y}}_{\mathrm{spl}}; \lambda\right) = \frac{1}{|A|^2} \sum_{i=1}^{n} \sum_{j>i}^{n} \frac{\pi(\mathbf{x}_i)\pi(\mathbf{x}_j) - \pi(\mathbf{x}_i, \mathbf{x}_j)}{\pi(\mathbf{x}_i, \mathbf{x}_j)} \left(\frac{\hat{e}_{(-i)}(\mathbf{x}_i)}{\pi(\mathbf{x}_i)} - \frac{\hat{e}_{(-j)}(\mathbf{x}_j)}{\pi(\mathbf{x}_j)} \right)^2 . \quad (8)$$

Then, the value of λ for PSRE is chosen to be

$$\lambda_s = \arg\min_{\lambda \geq 0} \hat{V}_{\mathrm{CV}}\left(\hat{\bar{Y}}_{\mathrm{spl}}; \lambda\right).$$

This criterion amounts to select the value of λ that minimizes the estimated sampling variance computed with the "leave-one-out" residuals.

It is worthwhile noting that it is not necessary to repeat the fitting to compute $\hat{e}_{(-i)}(\mathbf{x}_i)$, since for linear smoothers whose smoothing matrix has each row summing to one, as $\mathbf{S}_{\lambda s}$ in (6), it can be shown that

$$\hat{e}_{(-i)}(\mathbf{x}_i) = \frac{\hat{e}(\mathbf{x}_i)}{1 - S_{\lambda s}(i, i)},$$

where $S_{\lambda s}(i, i)$, is the (i, i)-th entry of the smoother matrix $\mathbf{S}_{\lambda s}$ (see Hastie and Tibshirani 1990, p. 47).

The rationale behind the above procedure can be summarized as follows. First, replacing $\hat{e}(\mathbf{x}_i)$ with $\hat{e}_{(-i)}(\mathbf{x}_i)$ in (7) preserves the estimator against the risk of overfitting. Second, as a function of λ, it allows choosing the value of $\hat{\bar{Y}}_{\mathrm{spl}}$ for which $\hat{V}_{\mathrm{CV}}\left(\hat{\bar{Y}}_{\mathrm{spl}}; \lambda\right)$ and, hopefully, the mean square error of $\hat{\bar{Y}}_{\mathrm{spl}}$, i.e. $\mathrm{MSE}_p\left(\hat{\bar{Y}}_{\mathrm{spl}}\right)$, attains its minimum value. In this respect, Opsomer and Miller (2005) proved that the design-based CV statistics they propose converges to the MSE of the local polynomial estimator under a suitable asymptotic framework and conditions on the sampling design and on the population.

3 The Block Kriging Predictor of the Mean

In Geostatistics the problem we are dealing with is managed with kriging, which is a model-based technique producing a predictor of the response variable which is unbiased and with minimum prediction variance under a second order stationary process. The population mean is estimated with the mean of the predicted values for all locations in the domain. The resulting estimator of the mean is called *block kriging predictor*. This estimator, which is essentially a weighted average of the response variable sample values, depends heavily on the coordinates of the sample points in $s = (\mathbf{x}_1, \ldots, \mathbf{x}_n)$. In fact, loosely speaking, the weights attached to observations $y(\mathbf{x}_1), \ldots, y(\mathbf{x}_n)$ are functions of the distances between the sample locations, as well as of the distances between sampled and unsampled locations in the domain (Cicchitelli and Montanari 1997).

The above features give to block kriging predictor the role of a natural benchmark for sampling strategies that make use of the coordinates of sample points within the design-based paradigm. In fact, the block kriging predictor is model unbiased and with the minimum prediction variance conditional on the selected sample; then, provided that the model holds true, it follows that jointly under the model and the sampling design it is unbiased and with the minimum prediction variance. For this reason, it makes sense comparing the performance of alternative estimators with that of block kriging predictor, even if it is evaluated from a repeated sampling point of view. In addition, studying the behaviour of this predictor in repeated sampling from a fixed population may be useful to explore its appropriateness as an estimator within the design-based framework. In this respect, McArthur (1987) compares kriging and design-based methods on simulated spatial data, concluding that kriging is design biased. Brus and de Gruijter (1997) present a simulation study where a stratified design coupled with the Horvitz–Thompson estimator is compared with the kriging predictor combined with systematic sampling. Their overall conclusion is that the kriging estimator is more efficient than the Horvitz–Thompson estimator for large sample size, but it often presents poor confidence interval coverage rates due to the fact that the kriging variance is not a good estimate of the sampling variance. Ver Hoef (2002) compares the kriging predictor with the sample mean in repeated simple random samples drawn from an artificial population. He shows that the kriging predictor is more efficient than the sample mean and gives valid confidence intervals.

We now give a technical sketch of the block kriging predictor for the population mean. Assuming a second order stationary stochastic process, it follows that $E_\xi[y(\mathbf{x})] = \mu$, where $E_\xi[\cdot]$ denotes expectation under the model, and that the covariance between $y(\mathbf{x})$ and $y(\mathbf{x} + \mathbf{h})$ is a function of $\mathbf{h}$ only, $C(\mathbf{h})$ (second-order stationarity). The covariance function is generally expressed by parsimonious models, under the assumption of isotropy. An important class of isotropic covariance functions is the Matérn family, which involves a three-parameter vector $\theta = (\sigma^2, \rho, \nu)$, where σ^2 is the variance, ρ is the range parameter (it controls how fast correlation decay with increasing distance) and ν is the smoothing parameter (it controls the smoothness of the resulting interpolating surface). The covariance is rarely known and must be estimated.

The block kriging, which is the (model) best linear unbiased predictor of the population mean, is given by (see Cicchitelli and Montanari 1997; Ver Hoef 2002)

$$\hat{\bar{Y}}_{kr} = \hat{\mu} + \mathbf{c}_s'\mathbf{V}_s^{-1}\left(\mathbf{y}_s - \mathbf{1}_s\hat{\mu}\right), \tag{9}$$

where $\hat{\mu} = \left(\mathbf{1}_s'\mathbf{V}_s^{-1}\mathbf{1}_s\right)^{-1}\mathbf{1}_s'\mathbf{V}_s^{-1}\mathbf{y}_s$ is the weighted least squares estimator of μ; $\mathbf{1}$ is the unit vector of proper size; $\mathbf{V}_s$ is the $n \times n$ dimensional matrix whose entries are the covariances between sample locations; $\mathbf{c}_s$ is the n-dimensional vector whose i-th entry is given by

$$c_i = \frac{1}{|A|}\int_A C(\mathbf{x} - \mathbf{x}_i)d\mathbf{x},$$

i.e., the mean covariance between $\mathbf{x}_i$ and any other location $\mathbf{x}$ in A.

The prediction variance is given by

$$V_\xi\left(\hat{\bar{Y}}_{kr}\right) = \sigma_{A,A}^2 - \mathbf{c}_s'\mathbf{V}_s^{-1}\mathbf{c}_s + d^2\left(\mathbf{1}_s'\mathbf{V}_s^{-1}\mathbf{1}_s\right)^{-1}, \tag{10}$$

where $V_\xi(\cdot)$ denotes variance under the model, $\sigma_{A,A}^2 = |A|^{-2}\int_A\int_A C(\mathbf{x} - \mathbf{x}')d\mathbf{x}d\mathbf{x}'$, and $d = 1 - \mathbf{1}'\mathbf{V}_s^{-1}\mathbf{c}_s$. An estimate of this variance is obtained replacing the integrals in (10) with numerical evaluations.

The properties of block kriging are conditional on the selected sample. Hence, provided that the model holds true, averaging across samples we get that $E_p E_\xi\left(\hat{\bar{Y}}_{kr} - \bar{Y}\right) = 0$ and $E_p E_\xi\left[\left(\hat{\bar{Y}}_{kr} - \bar{Y}\right)^2\right] E_p\left[V_\xi\left(\hat{\bar{Y}}_{kr}\right)\right]$ attains the minimum among all linear predictors that are model unbiased.

Of course, we are interested in design-based procedures since they are model free, in the sense that the validity of inference based on them does not depend on model assumptions.

4   Simulation Results

A simulation study aimed at comparing estimator $\hat{\bar{Y}}_{spl}$ with the block kriging predictor $\hat{\bar{Y}}_{kr}$ has been carried out. We considered several artificial populations on the unit square with different degrees of smoothness. For the sake of brevity, we report only on two of them, depicted in Fig. 1 and given by the following functions:

Population A

$$y(\mathbf{x}) = 13.333 \times \left\{5[\sin(x_1)]^2 + 5[\cos(x_2)]^2 + 5x_1\right\}, \quad 0 < x_1 < 1, \ 0 < x_2 < 1.$$

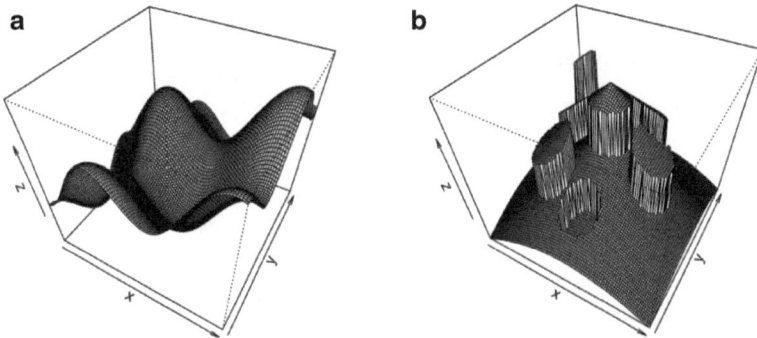

Fig. 1 Surface $y(\mathbf{x})$ of population A (**a**) and population B (**b**)

Population B

$$y(\mathbf{x}) = 335.96 \times \Big[h_{0.5,0.5,0.5,1}(\mathbf{x}) + g_{0.3,0.3,0.1}(\mathbf{x}) - g_{0.5,0.2,0.1}(\mathbf{x}) + g_{0.5,0.5,0.5,1}(\mathbf{x})$$
$$+ g_{0.3,0.7,0.1}(\mathbf{x}) + g_{0.6,0.7,0.1}(\mathbf{x}) + g_{0.25,0.6,0.1}(\mathbf{x})$$
$$+ g_{0.75,0.45,0.1}(\mathbf{x}) - g_{0.7,0.6,0.1}(\mathbf{x}) \Big], \quad 0 < x_1 < 1, \ 0 < x_2 < 1,$$

where

$$g_{a,b,r}(\mathbf{x}) = \frac{1}{2} \cos \Big\{ \frac{\pi}{2} \big[(x_1 - a)^2 + (x_2 - b)^2 \big] \Big\} I_{B_r(a,b)}(\mathbf{x})$$
$$h_{a,b,r,q}(\mathbf{x}) = 2^{1/q} \Big\{ r^q - \big[(x_1 - a)^2 + (x_2 - b)^2 \big]^{1/q} \Big\},$$

and $I_{B_r(a,b)}(\mathbf{x})$ is the indicator function of belonging to the disk of radius r and centre (a, b).

The population mean of both populations is equal to 100.0. The variances are 20,545 and 1,728, respectively, for population A and population B. The corresponding coefficients of variation are 143.3 and 41.6 %, respectively.

Three sampling strategies on the unit square are considered: (a) uniform random sampling, (b) stratified uniform random sampling with two units per stratum, (c) stratified uniform random sampling with one unit per stratum. The sample size is fixed to be $n = 100$ and $n = 256$. The strata are built superimposing a grid of 10×10 smaller squares for $n = 100$ and 16×16 for $n = 256$. In the case (b), pairs of adjacent strata are collapsed along the rows to obtain 128 strata overall. In the case (c), the strata are considered collapsed in the same way for variance estimation only.

One thousand random samples are selected for each sampling strategy and sample size and for each of them estimators $\hat{\bar{Y}}_{spl}$ and $\hat{\bar{Y}}_{kr}$ are computed (for the sake of brevity, we put $\hat{\bar{Y}}_1 = \hat{\bar{Y}}_{spl}$ and $\hat{\bar{Y}}_2 = \hat{\bar{Y}}_{kr}$). Monte Carlo estimates of the following quantities are at last obtained:

- Per cent bias with respect to the mean

$$R = \left[E_{MC}\left(\hat{\bar{Y}}_i\right) - \bar{Y} \right] / \bar{Y}, \quad i = 1, 2.$$

- Relative efficiency (with respect to the sample mean, $\bar{y}$)

$$\text{Eff}_{MC}\left(\hat{\bar{Y}}_i\right) = \frac{\text{MSE}_{MC}(\bar{y})}{\text{MSE}_{MC}\left(\hat{\bar{Y}}_i\right)}, \quad i = 1, 2.$$

- Variance to mean square error ratio

$$R_{var/mse} = \frac{E_{MC}\left[\hat{V}\left(\hat{\bar{Y}}_i\right)\right]}{\text{MSE}_{MC}\left(\hat{\bar{Y}}_i\right)}, \quad i = 1, 2.$$

- Confidence interval width and coverage for the 95 % nominal level based on the normal approximation.

The number of knots appearing in model (1) is fixed to be equal to the sample size and their locations are established using the algorithm of Nychka et al. (1998).

In each sample, the value of $\hat{\bar{Y}}_{spl}$ is computed with λ_s, obtained by minimizing the design-based CV criterion (8). The variance of $\hat{\bar{Y}}_{spl}$ is estimated with (8), apart when uniform random sampling is employed. In the latter case, it is common to compensate for the estimation of the residual $\hat{e}_{(-i)}(\mathbf{x}_i)$, adjusting for the degrees of freedom. Inspired by this argument, we adopt as variance estimator for uniform random sampling the following version

$$\hat{V}_{CV}^m\left(\hat{\bar{Y}}_{spl}; \lambda\right) = \frac{n-1}{n - r(\lambda)} \hat{V}_{CV}\left(\hat{\bar{Y}}_{spl}; \lambda\right),$$

where $r(\lambda)$ is the number of degrees of freedom corresponding to the chosen value of λ.

The kriging predictor is computed assuming as covariance function the following isotropic exponential model

$$C(\mathbf{h}) = \begin{cases} \theta_1 + \theta_2, & \mathbf{h} = \mathbf{0}, \\ \theta_2 \exp\left(-\|\mathbf{h}\|/\theta_3\right), & \mathbf{h} \neq \mathbf{0}. \end{cases}$$

Its parameters are estimated for each sample by means of the restricted maximum likelihood (RML).

The main results of the simulation are shown in Table 1 for uniform random sampling, in Table 2 for the two units per stratum design, and in Table 3 for the one unit per stratum design.

Table 1 Performances of estimators under uniform random sampling

Estimator	Population A		Population B	
	$\hat{\bar{Y}}_{spl}$	$\hat{\bar{Y}}_{kr}$	$\hat{\bar{Y}}_{spl}$	$\hat{\bar{Y}}_{kr}$
	$n = 100$			
Average of r	73.6	–	53.7	–
Bias ratio (%)	−0.02	−0.03	−0.15	0.92
$\mathrm{MSE}\left(\hat{\bar{Y}}_i\right)$	0.2138	0.6651	43.84	40.28
$\mathrm{Eff}_{MC}\left(\hat{\bar{Y}}_i\right)$	78.0	25.1	4.66	5.08
$R_{var/mse}$	1.82	1.88	1.59	1.07
Average width	2.37	4.38	32.5	25.6
Coverage (%)	96.3	99.0	97.0	94.7
	$n = 256$			
Average of r	179.0	–	108.3	–
Bias ratio (%)	−0.05	−0.02	−0.02	0.08
$\mathrm{MSE}\left(\hat{\bar{Y}}_i\right)$	0.0080	0.0285	9.298	9.208
$\mathrm{Eff}_{MC}\left(\hat{\bar{Y}}_i\right)$	835.7	235.2	8.89	8.98
$R_{var/mse}$	1.77	4.60	1.22	0.98
Average width	0.46	1.42	13.22	11.73
Coverage (%)	98.0	100.0	95.5	95.5

Table 2 Performances of estimators under two units per stratum design

Estimator	Population A		Population B	
	$\hat{\bar{Y}}_{spl}$	$\hat{\bar{Y}}_{kr}$	$\hat{\bar{Y}}_{spl}$	$\hat{\bar{Y}}_{kr}$
	$n = 100$			
Average of r	78.9	–	29.0	–
Bias ratio (%)	0.06	−0.04	0.11	0.80
$\mathrm{MSE}\left(\hat{\bar{Y}}_i\right)$	0.0576	0.1229	28.95	21.80
$\mathrm{Eff}_{MC}\left(\hat{\bar{Y}}_i\right)$	29.9	17.6	1.47	1.96
$R_{var/mse}$	1.60	7.07	1.11	1.40
Average width	1.30	3.65	21.8	21.5
Coverage (%)	96.9	100.0	93.0	96.8
	$n = 256$			
Average of r	186.1	–	70.2	–
Bias ratio (%)	−0.01	−0.01	−0.19	−0.02
$\mathrm{MSE}\left(\hat{\bar{Y}}_i\right)$	0.0033	0.0081	5.964	5.756
$\mathrm{Eff}_{MC}\left(\hat{\bar{Y}}_i\right)$	101.9	41.5	1.99	2.06
$R_{var/mse}$	1.54	11.1	1.29	1.06
Average width	0.275	1.17	10.8	9.6
Coverage (%)	96.8	100.0	95.5	93.0

Table 3 Performances of estimators under one unit per stratum design

	Population A		Population B	
Estimator	$\hat{\bar{Y}}_{\text{spl}}$	$\hat{\bar{Y}}_{\text{kr}}$	$\hat{\bar{Y}}_{\text{spl}}$	$\hat{\bar{Y}}_{\text{kr}}$
	$n = 100$			
Average of r	78.6	–	32.2	–
Bias ratio (%)	−0.01	0.00	−0.08	0.19
$\text{MSE}\left(\hat{\bar{Y}}_i\right)$	0.0389	0.0475	17.29	15.04
$\text{Eff}_{\text{MC}}\left(\hat{\bar{Y}}_i\right)$	25.2	20.6	1.39	1.59
$R_{\text{var/mse}}$	3.30	14.32	1.93	1.73
Average width	1.374	3.234	22.47	19.94
Coverage (%)	99.8	100.0	98.3	98.3
	$n = 256$			
Average of r	187.2	–	111.6	–
Bias ratio (%)	0.00	0.01	0.01	0.34
$\text{MSE}\left(\hat{\bar{Y}}_i\right)$	0.0031	0.0077	5.248	5.092
$\text{Eff}_{\text{MC}}\left(\hat{\bar{Y}}_i\right)$	121.6	49.0	1.85	1.91
$R_{\text{var/mse}}$	1.23	12.35	1.26	1.28
Average width	0.238	1.208	10.05	10.01
Coverage (%)	94.6	100.0	96.5	98.0

In the tables we report the across samples average number of the degrees of freedom; the per cent relative bias; the ratio between the mean squared error of the sample mean and that of the estimators $\hat{\bar{Y}}_{\text{spl}}$ and $\hat{\bar{Y}}_{\text{kr}}$; the ratio between the average of variance estimates and the mean squared error of the estimators; the average width of the confidence intervals; the 95 % confidence interval per cent coverage.

In all cases, the two estimators behave as nearly unbiased; the relative bias is always below 0.5 %, apart for the kriging predictor in the uniform random sampling of size 100.

Looking at the MSE of the estimators, for population A, which is more regular than population B, the efficiency of $\hat{\bar{Y}}_{\text{spl}}$ is much higher than that of $\hat{\bar{Y}}_{\text{kr}}$, and both estimators are far more efficient than the sample mean. The efficiency with respect to the sample mean increases with the sample size and decreases passing to more efficient sampling design, as the stratified ones. In fact, the sample mean estimator interpolates the response variable with the stratum sample mean and some of the efficiency is captured by the design. However, the working models upon which PSRE and the kriging predictor are based provide some extra efficiency with respect to the sampling design. In the case of population B, which is sharper, $\hat{\bar{Y}}_{\text{kr}}$ is somewhat more efficient than $\hat{\bar{Y}}_{\text{spl}}$. Note that the higher the degree of smoothness of the response variable, the higher is the number of degrees of freedom selected for $\hat{\bar{Y}}_{\text{spl}}$.

The high number of degrees of freedom selected for PSRE prevents the use of standard variance estimator (7), which would be dramatically downward biased. The alternatives based on the "leave-one-out" residuals explored in this simulation are generally positively biased and overestimate the MSE of the estimator. But, the variance estimator of $\hat{\bar{Y}}_{kr}$ can be several times bigger than the MSE for population A, revealing a possible inadequacy of the second order stationary model. As a consequence, confidence intervals based on the normal approximation are conservative and unnecessarily large. In the case of one unit per stratum design, which is not a measurable design, the collapsed strata technique for estimating the variance results in a severely positively biased estimator and this is particularly clear in the case of population A and sample size equal to 100.

In summary, the efficiency of the penalized spline regression estimator is close to that of the kriging and much higher in the case of population A, when the number of degrees of freedom of the spline regression model is large. In all cases, valid, even if conservative, confidence intervals are obtained with $\hat{\bar{Y}}_{spl}$.

5   Conclusions

This paper starts from the results of Cicchitelli and Montanari (2012), where a model-assisted penalized spline regression estimator (PSRE) of a spatial population mean is introduced and it is proved that this estimator is design-consistent and approximately unbiased. The main idea is to capture the spatial pattern in the data by assuming that the target population comes from a super-population described by a penalized spline regression model. In this approach, the role of the model in strengthening the efficiency of the design-based estimation is similar to that of the covariance function in the block kriging prediction procedure.

In the above paper, the smoothing parameter that determines the behaviour of the estimator is treated as a fixed quantity and the issue of how to best select its value is not addressed. Another open problem is the poor performance of the standard design-consistent variance estimator when the number of degrees of freedom of the smoother is relatively high with respect to the sample size.

The two issues are addressed in this paper making use of the "leave-one-out" cross-validation criterion. We assign to the tuning parameter λ the value that minimizes the sampling variance computed with the "leave-one-out" residuals. This solution follows heuristically from that of Opsomer and Miller (2005), who propose a design-based cross-validation criterion for selecting the bandwidth of a local polynomial regression estimator.

With regard to PSRE, the evidence coming from the simulation study presented in this article can be summarized as follows.

In uniform random sampling and in stratified random sampling with two units per stratum, the cross-validation criterion for selecting the smoothing parameter and the variance estimator used here seems to work quite well. This procedure provides valid, even if conservative, confidence intervals. PSRE strongly outperforms block kriging predictor when the response variable is given by a smooth surface, while kriging seems to work a little better when the population is given by a sharp surface. In the first situation, the number of degrees of freedom of PSRE tends to be very high.

In stratified sampling with one observation per stratum, PSRE presents, as expected, a mean squared error which continues to be considerably smaller than that obtained with samples of the same size drawn with simple random sampling. But the estimator of variance, based on collapsing contiguous pairs of strata, may seriously overestimate the true variance.

As an overall assessment, the penalized spline regression estimator performs well or better than block kriging predictor, in contrast to the findings of other studies where design-based estimation strategies and block kriging are compared (see, for example, Brus and de Gruijter 1997; Ver Hoef 2002). This is due to the fact that in the former, differently from what is done in the above papers, the spatial structure of the population is conveyed, at the estimation stage, into the estimator through the spline regression assisting model. In other words, it can be said that sampling theory and geostatistics are reconciled, thanks to the duality between first-order and second-order modelling in geostatistics, as mentioned in Shaddick and Zidek (2012).

We must point out that the technique for selecting the smoothing parameter and for estimating the variance mimics the procedure proposed by Opsomer and Miller (2005), in a finite population sampling context, for bandwidth selection in local polynomial regression estimation; in the latter important regularity conditions were imposed on the sampling method and on the population. Our results rely mainly on empirical evidences and need to be supported by appropriate theoretical results. In particular, it has to be established if the design consistency of the estimator proved in Cicchitelli and Montanari (2012) for a fixed value of the smoothing parameter continues to hold in the new context, where a data-driven technique is used for selecting it.

A second open problem is the search for an alternative estimator of the variance in the case of stratified sampling with one observation per stratum, which appears to be one of the most efficient way to draw samples from a spatial population, as various papers have shown (Barabesi et al. 2012; Stevens 1997).

Of course, the approach can be generalized in several ways. For example, further covariates may be added to the assisting model, parametrically and non-parametrically. In such a case, the penalized spline part of the model accounts for the spatial covariance between residuals from the added covariates. It can be also shown that model (1) can be written as a mixed model and standard techniques in that field might be useful for estimation and inference.

Another area for future research is the use of the function $\hat{y}(\mathbf{x})$, given in (2), as a design-based estimator of the population surface $y(\mathbf{x})$ at point $\mathbf{x}$. Assuming that the function $y(\mathbf{x})$ is continuous (or integrable), as it is done within environmental monitoring, forestry, geology, etc. (see, for example, Gregoire and Valentine 2008; Barabesi et al. 2012), the consistency of $\hat{y}(\mathbf{x})$ can be conjectured on the basis of the available asymptotic results concerning the spline regression function. These results, coupled with an appropriate design-based variance estimator measuring the degree of uncertainty of the estimate, could allow the use of $\hat{y}(\mathbf{x})$ to estimate the population value $y(\mathbf{x})$ and to draw maps, as the kriging predictor does in the model-based context.

References

Barabesi, L., Franceschi, S., Marcheselli, M.: Properties of design-based estimation under stratified spatial sampling with application to canopy coverage estimation. Ann. Appl. Stat. **6**, 210–228 (2012)

Brus, D.: Using regression models in design-based estimation of spatial means of soil properties. Eur. J. Soil Sci. **51**, 159–172 (2000)

Brus, D., de Gruijter, J.: Random sampling or geostatistical modeling? Choosing between design-based and model-based strategies for soil (with discussion). Geoderma **80**, 1–59 (1997)

Cicchitelli, G., Montanari, G.E.: The kriging predictor for spatial finite population inference. Metron **55**, 41–57 (1997)

Cicchitelli, G., Montanari, G.E.: Design-based estimation of a spatial population mean. Int. Stat. Rev. **80**, 111–126 (2012)

Cordy, C.B.: An extension of the Horvitz–Thompson theorem to point sampling from a continuous universe. Stat. Probab. Lett. **18**, 353–362 (1993)

Gregoire, T.G., Valentine, H.T.: Sampling Strategies for Natural Resources and the Environment. Chapman & Hall, New York (2008)

Hastie, T.J., Tibshirani, R.J.: Generalized Additive Models. Chapman & Hall, New York (1990)

McArthur, R.D.: An evaluation of sample designs for estimating a locally concentrated pollutant. Commun. Stat. Simulat. Comput. **16**, 735–759 (1987)

Nychka, D., Haaland, P., O'Connel, M., Ellner, S.: FUNFITS. Data analysis and statistical tools for estimating functions. In: Nychka, D., Piegorsch, W.W., Cox, L.H. (eds.) Case Study in Environmental Statistics, Lecture Notes in Statistics, vol. 132, pp. 159–179. Springer, New York (1998)

Opsomer, J.D., Miller, C.P.: Selecting the amount of smoothing in nonparametric regression estimation for complex surveys. Nonparametric Stat. **17**, 593–611 (2005)

Ruppert, D., Wand, M.P., Carroll, R.J.: Semiparametric Regression. Cambridge University Press, Cambridge (2003)

Särndal, C.E., Svensson, B., Wretman, J.: Model Assisted Survey Sampling. Springer, New York (1992)

Shaddick, G., Zidek, J.V.: Unbiasing estimates from preferentially sampled spatial data. Technical report nr. 268, Department of Statistics, The University of British Columbia (2012)

Stevens, D.L.: Variable density grid-based sampling designs for continuous spatial populations. Environmetrics **8**, 167–195 (1997)

Stevens, D.L., Jensen, S.F.: Sample design, execution, and analysis for wetland assessment. Wetlands **27**(3), 515–523 (2007)

Stevens, D.L., Olsen, A.R.: Spatially balanced sampling of natural resources. J. Am. Stat. Assoc. **99**, 262–278 (2004)

Ver Hoef, J.: Sampling and geostatistics for spatial data. Ecoscience **9**(2), 152–161 (2002)

Optimal Regression Estimator for Stratified Two-Stage Sampling

Nuanpan Nangsue and Yves G. Berger

Abstract Regression estimators are often used in survey sampling for point estimation. We propose a new regression estimator based upon the optimal estimator proposed by Berger et al., (2003). The proposed estimator can be used for stratified two-stage sampling designs when the sampling fraction is negligible and the primary sampling units (PSU) are selected with unequal probabilities. For example, this is the case for self-weighted two-stage designs. We assume that we have auxiliary variables available for the secondary sampling units (SSU) and the primary sampling units (PSU). We propose to use an ultimate cluster approach to estimate the regression coefficient of the regression estimator. Estevao and Särndal (2006) proposed a regression estimator for two-stage sampling. This estimator will be compared with the proposed estimator through a simulation study. We will show that the proposed estimator is more accurate that the Estevao and Särndal (2006) estimator when the strata are homogeneous.

Keywords Design-based approach • Horvitz–Thompson estimator • Stratification • Ultimate cluster approach • Unequal inclusion probabilities

1 Introduction

Regression estimators are used for design-based estimation of population totals in survey sampling. Berger et al. (2003) proposed an asymptotically optimal regression estimator based on the Montanari (1987) estimator. Tan (2013) proposed an optimal

N. Nangsue • Y.G. Berger (✉)
University of Southampton, Southampton, SO17 1BJ, UK
e-mail: y.g.berger@soton.ac.uk

F. Mecatti et al. (eds.), *Contributions to Sampling Statistics*, Contributions to Statistics,
DOI 10.1007/978-3-319-05320-2_11,
© Springer International Publishing Switzerland 2014

regression estimator which is a particular case of the estimator proposed by Berger et al. (2003). In the rest of this paper, the concept of optimality should be understood as asymptotic optimality. In this paper, we propose to extend the estimator proposed by Berger et al. (2003) for two-stage sampling. Estevao and Särndal (2006) proposed a regression estimator which uses auxiliary information available at both primary sampling units (PSU) and secondary sampling units (SSU) level. The proposed estimator will be compared with Estevao and Särndal (2006) regression estimator via a simulation study.

Let N be the number of PSU in a population U, and M_i the number of SSU in PSU i where $i = 1, 2, \ldots, N$. We consider that the population of PSU is divided into H strata: $U_1, \ldots, U_h, \ldots, U_H$; where $\cup_{h=1}^{H} U_h = U$. Within the stratum U_h, we have N_h PSU, where $N = \sum_{h=1}^{H} N_h$. A sample s_h of n_h PSU are selected without replacement within stratum h. The overall sample of PSU is given by $s = \cup_{h=1}^{H} s_h$. Let π_i be the inclusion probability of the i-th PSU. Therefore, the number of sampled PSU is given by $n = \sum_{h=1}^{H} n_h$. We assume that n_h/N_h is negligible and that H is asymptotically bounded.

A sample s_i of m_i SSU is selected from the PSU i sampled. Let $\pi_{j|i}$ be the conditional inclusion probability of the j-th SSU given the selection of the i-th PSU. The number of SSU selected is given by $m = \sum_{i \in s} m_i$. The sample of SSU is denoted by s^*.

The aim is to estimate the unknown population total t_y given by

$$t_y = \sum_{i \in U} \sum_{j \in \mathrm{PSU}_i} y_{ij},$$

where y_{ij} is the value of the variable of interest for the j-th SSU from the i-th PSU denoted by PSU_i. The Horvitz and Thompson (1952) estimator (see also Narain 1951) of t_y is given by

$$\hat{t}_{y\pi} = \sum_{i \in s} \sum_{j \in s_i} \breve{y}_{ij},$$

where $\breve{y}_{ij} = y_{ij}/\pi_j^*$; where $\pi_j^* = \pi_{j|i} \pi_i$. It can be easily shown that this estimator is design unbiased.

In Sect. 2, we define the regression estimator proposed by Estevao and Särndal (2006). In Sect. 3, we define the optimal regression estimator proposed by Montanari (1987) and Berger et al. (2003) for single stage design. In Sect. 4, we propose to extend the optimal estimator proposed by Berger et al. (2003) for two-stage sampling. In Sect. 5, we compare the proposed estimator with the Estevao and Särndal (2006) estimator via a simulation study.

2 Regression Estimator Under Two-Stage Sampling

With two-stage surveys, we often have auxiliary information available at PSU and SSU level. Suppose that we have q auxiliary variables available at PSU level (information-Z) and p auxiliary variables available at PSU level (information-X). Let x_{ij} be a $p \times 1$ vector containing the auxiliary variables of the j-th SSU of the i-th PSU. Let z_i be the $q \times 1$ vector containing the auxiliary variables of the i-th PSU level. Estevao and Särndal (2006) suggest to use the following SSU level variable defined by

$$z_{ij} = \frac{z_i}{M_i}.$$

We assume that the following totals are known.

$$t_x = \sum_{i=1}^{N} \sum_{j=1}^{M_i} x_{ij},$$

$$t_z = \sum_{i=1}^{N} \sum_{j=1}^{M_i} z_{ij} = \sum_{i=1}^{N} z_i.$$

Estevao and Särndal (2006) proposed the following regression estimator which uses the information-X and the information-Z.

$$\hat{t}_{\mathrm{reg}}^{(\mathrm{ES})} = \hat{t}_{y\pi} + (t_{xz} - \hat{t}_{xz\pi})^{\top} \hat{\beta}_{xz}, \tag{1}$$

where $t_{xz} = (t_x^{\top}, t_z^{\top})^{\top}$, $\hat{t}_{xz} = (\hat{t}_{x\pi}^{\top}, \hat{t}_{z\pi}^{\top})^{\top}$,

$$t_x = \sum_{i \in U} \sum_{j \in \mathrm{PSU}_i} x_{ij},$$

$$t_z = \sum_{i \in U} \sum_{j \in \mathrm{PSU}_i} z_{ij},$$

$$\hat{t}_{x\pi} = \sum_{i \in s} \sum_{j \in s_i} \check{x}_{ij}, \quad \text{where } \check{x}_{ij} = \frac{x_{ij}}{\pi_j^*},$$

$$\hat{t}_{z\pi} = \sum_{i \in s} \sum_{j \in s_i} \check{z}_{ij}, \quad \text{where } \check{z}_{ij} = \frac{z_{ij}}{\pi_j^*},$$

$$\hat{\beta}_{xz} = (W_s^{*\top} \Sigma_s W_s^*)^{-1} W_s^{*\top} \Sigma_s^* Y_s^*,$$

$$W_s^* = [X_s^* \vert Z_s^*],$$

$$\Sigma_s^* = diag\{\pi_\ell^{*-1} : \ell \in s^*\},$$

X_s^* denotes the $m \times p$ matrix containing the values of the auxiliary variable x_{ij} for the sampled SSU, Z_s^* denotes the $m \times q$ matrix containing the values of the auxiliary variable z_{ij} for the sampled SSU and Y_s^* denotes the $m \times 1$ vector containing the values of variable of interest y_{ij} for the sampled SSU. Note that X_s^*, Z_s^*, Y_s^* and Σ_s^* contain the values at SSU level.

3 Asymptotically Optimal Regression Estimator

Montanari (1987) considered the following random variable

$$\tilde{t}_M = \hat{t}_{y\pi} + (t_{xz} - \hat{t}_{xz\pi})' \beta_{\text{opt}},$$

where

$$\beta_M = var(\hat{t}_{xz\pi})^{-1} cov(\hat{t}_{xz\pi}, \hat{t}_{y\pi}) \tag{2}$$

is a population regression parameter and $var(\cdot)$ and $cov(\cdot, \cdot)$ denote respectively the design-based variance and covariance operators. Montanari (1987) showed that $\tilde{t}_M$ is optimal because the expectation of $\tilde{t}_M$ equal t_y and the variance of $\tilde{t}_M$ is minimal. Montanari (1987) proposed to predict $\tilde{Y}_M$ by substituting β_{opt} by its consistent estimator. For two-stage sampling designs, this estimator is very complex to implement because it requires the joint-inclusion probabilities. Berger et al. (2003) showed that under single stage design a consistent estimator of β_{opt} can be obtained by including the stratification variable into the regression estimator (see also Tan 2013). This estimator does not involve joint-inclusion probabilities. In a simulation study, Berger et al. (2003) showed that this estimator may be more accurate than the standard generalised regression estimator. This simulation is limited to single stage sampling design. In Sect. 4, we propose to extend the estimator proposed by Berger et al. (2003) for two-stage sampling.

4 Proposed Regression Estimator for Two-Stage Sampling

We propose to extend the optimal estimator proposed by Berger et al. (2003) for two-stage sampling. The proposed approach consists in using an ultimate cluster approach to estimate the regression coefficient (2). We will see that it is necessary to incorporate the stratification variables into the estimation of the regression coefficient. The ultimate cluster approach is a common method for variance estimation in complex survey designs which consists in treating the PSU as sampling units when estimating variances and covariances. This approach was proposed by Hansen et al. (1953).

Let $\boldsymbol{q}_i = (q_{1i}, \ldots, q_{hi}, \ldots, \check{q}_{Hi})^\top$ $(i \in s)$ be the stratification variables, where

$$q_{hi} = \begin{cases} \pi_i & \text{if } i \in U_h, \\ 0 & \text{otherwise;} \end{cases}$$

is the PSU level variable specifying the h-th stratum.

The *proposed optimal regression estimator* is given by

$$\hat{t}_{\text{reg}}^{(\text{opt})} = \hat{t}_{y\pi} + (\boldsymbol{t}_{xzq} - \hat{\boldsymbol{t}}_{xzq\pi})^\top \hat{\boldsymbol{\beta}}_{\text{opt}}, \tag{3}$$

where $\boldsymbol{t}_{xzq\pi} = (\boldsymbol{t}_{x\pi}^\top, \boldsymbol{t}_{z\pi}^\top, \boldsymbol{t}_q^\top)^\top$, $\hat{\boldsymbol{t}}_{xzq\pi} = (\hat{\boldsymbol{t}}_{x\pi}^\top, \hat{\boldsymbol{t}}_{z\pi}^\top, \hat{\boldsymbol{t}}_{q\pi}^\top)^\top$,

$$\boldsymbol{t}_q = \sum_{i \in U} \boldsymbol{q}_i,$$

$$\hat{\boldsymbol{t}}_{q\pi} = \sum_{i \in s} \frac{\boldsymbol{q}_i}{\pi_i},$$

$$\hat{\boldsymbol{\beta}}_{\text{opt}} = (\check{\boldsymbol{\Gamma}}_s^\top \check{\boldsymbol{C}}_s \check{\boldsymbol{\Gamma}}_s)^{-1} \check{\boldsymbol{\Gamma}}_s^\top \check{\boldsymbol{C}}_s \check{\boldsymbol{Y}}_s, \tag{4}$$

$$\check{\boldsymbol{\Gamma}}_s = \left[\check{\boldsymbol{X}}_s \vdots \check{\boldsymbol{Z}}_s \vdots \check{\boldsymbol{Q}}_s \right],$$

$$\check{\boldsymbol{C}}_s = \boldsymbol{diag}\{1 - \pi_i : i \in s\},$$

where $\check{\boldsymbol{X}}_s$ denote the $n \times p$ matrix which contains the estimators $\check{\boldsymbol{x}}_i$ of the PSU total divided by π_i, $\check{\boldsymbol{Z}}_s$ denote the $n \times q$ matrix which contains the estimators $\check{\boldsymbol{z}}_i$ of the PSU total divided by π_i, $\check{\boldsymbol{Q}}_s$ denote the $n \times H$ matrix which contains the stratification variables $\check{\boldsymbol{q}}_i$ and $\check{\boldsymbol{Y}}_s$ denote the $n \times 1$ vector which contains the estimators $\check{y}_i$ of the PSU total of the variables of interest. The PSU total estimators $\check{\boldsymbol{x}}_i$, $\check{z}_i$ and $\check{y}_i$ are given by

$$\check{\boldsymbol{x}}_i = \sum_{j \in s_i} \frac{x_{ij}}{\pi_j^*}, \quad \check{z}_i = \sum_{j \in s_i} \frac{z_{ij}}{\pi_j^*}, \quad \check{y}_i = \sum_{j \in s_i} \frac{y_{ij}}{\pi_j^*}.$$

Note that $\check{\boldsymbol{\Gamma}}_s$ is a $n \times (p+q+H)$ matrix. The matrix $\check{\boldsymbol{C}}_s$ contains the finite population corrections $1 - \pi_i$. Note that as $\check{\boldsymbol{X}}_s$, $\check{\boldsymbol{Z}}_s$, $\check{\boldsymbol{Q}}_s$, $\check{\boldsymbol{Y}}_s$ and $\check{\boldsymbol{C}}_s$ contain information at PSU level. The regression coefficient $\hat{\boldsymbol{\beta}}_{\text{opt}}$ is estimated at PSU level.

We have that $\hat{\boldsymbol{t}}_{q\pi} = \boldsymbol{t}_q = \boldsymbol{n}$, where $\boldsymbol{n} = (n_1, \ldots, n_h, \ldots, n_H)^\top$ is the vector of the strata sample sizes. Thus, the Horvitz and Thompson (1952) estimator $\hat{\boldsymbol{t}}_{q\pi}$ does not have any sampling error. Hence, the last H stratification component of $\boldsymbol{t}_{xzq} - \hat{\boldsymbol{t}}_{xzq\pi}$ are equal to zero, as $\boldsymbol{t}_{xzq} - \hat{\boldsymbol{t}}_{xzq\pi} = (\boldsymbol{t}_x^\top - \hat{\boldsymbol{t}}_{x\pi}^\top, \boldsymbol{t}_z^\top - \hat{\boldsymbol{t}}_{z\pi}^\top, \boldsymbol{0}^\top)^\top$. Thus, the last H component of $\hat{\boldsymbol{\beta}}_{\text{opt}}$ do not have any contribution in (3). Nevertheless, the regression coefficient depends on the stratification variables which need to be incorporated into the estimation of the regression coefficient.

When all the PSU contain one unit, we have a single stage sampling design. In this situation, it can be easily shown that (3) reduces to the optimal regression estimator proposed by Berger et al. (2003). The regression coefficient $\hat{\boldsymbol{\beta}}_{\text{opt}}$ is given by the inverse of the variance matrix, $\check{\boldsymbol{\Gamma}}_s^{\top} \check{\boldsymbol{C}}_s \check{\boldsymbol{\Gamma}}$, multiplied by the covariance matrix $\check{\boldsymbol{\Gamma}}_s^{\top} \check{\boldsymbol{C}}_s \check{\boldsymbol{Y}}_s$. When the sampling fraction is negligible, these variances and covariances are consistent estimator of the variances and covariances in (2). Berger et al. (2003) proved this results for single stage sampling design. By using a ultimate cluster approach, this result can be generalised and yield to the estimator (3).

5 Simulation

We generate M_i SSU, where the value of M_i is given by

$$M_i = aW_i + b,$$

with $W_i = (i/N)^{\alpha} + \alpha^{-1}$ (Deville 1997; Berger 2005). The quantity α controls the variation between the M_i. A large value for α implies a skewed distribution for the M_i. We use $\alpha = 4$. The values a and b are used to control the minimum and the maximum values of M_i.

The strata ($H = 3$) are constructed in two different ways: (a) "stratification by PSU size" where the PSU are sorted according to M_i. The first N_1 PSU are grouped together, the next N_2 PSU are grouped together and so on. This way PSU of similar sizes are grouped together within the same strata. In this case, the strata are homogeneous. (b) "random stratification" where the PSU are allocated randomly within stratum. In this case, the strata are heterogeneous.

The value of the variable of interest y_{ij} are generated from the following multilevel model:

$$y_{ij} = x_{ij}'\beta + z_i'\gamma + \epsilon_{1i} + \epsilon_{2ij},$$

with $x_{ij} \sim N(20, 1), z_i \sim N(0, 1)$,

$$\epsilon_{1i} \sim \begin{cases} N(-10, 1) & \text{if } i \in U_1, \\ N(0, 1) & \text{if } i \in U_2, \\ N(10, 1) & \text{if } i \in U_3 \end{cases}$$

and $\epsilon_{2ij} \sim N(0, \sigma_{\epsilon_{2ij}}^2)$, where $\sigma_{\epsilon_{2ij}}^2 = (1 - \rho)\rho^{-1}\sigma_{\epsilon_{1i}}^2$, $\rho = 0.1$ or 0.4, and $\beta_0 = \gamma_0 = 0$. ρ is the intra-cluster correlation. Several values of β and γ will be considered.

Table 1 Relative bias (%) $N_1 = 3,000$, $N_2 = 2,000$, $N_3 = 1,000$, $H = 3$

		r_{yx}	r_{yz}	HT	Estevao and Särndal			Proposed		
					X	Z	XZ	XQ	ZQ	XZQ
$\rho = 0.1$	Stratification by	0.2	0.2	0.36	0.35	0.35	0.25	0.35	0.25	0.24
	PSU size	0.2	0.7	1.06	1.05	0.66	0.58	1.06	0.52	0.51
		0.7	0.2	0.12	0.08	0.27	0.06	0.08	0.10	0.05
		0.7	0.7	0.31	0.30	0.33	0.16	0.30	0.17	0.14
	Random	0.2	0.2	0.36	0.34	0.46	0.24	0.34	0.30	0.29
	stratification	0.2	0.7	0.99	0.98	0.62	0.46	0.99	0.76	0.75
		0.7	0.2	0.12	0.09	0.37	0.05	0.09	0.10	0.07
		0.7	0.7	0.28	0.26	0.40	0.12	0.26	0.21	0.20
$\rho = 0.4$	Stratification by	0.2	0.2	0.34	0.33	0.33	0.22	0.33	0.22	0.20
	PSU size	0.2	0.7	1.06	1.05	0.66	0.58	1.06	0.52	0.51
		0.7	0.2	0.12	0.08	0.27	0.05	0.08	0.09	0.05
		0.7	0.7	0.31	0.30	0.33	0.16	0.30	0.17	0.14
	Random	0.2	0.2	0.33	0.32	0.45	0.21	0.32	0.28	0.26
	stratification	0.2	0.7	0.99	0.98	0.62	0.45	0.98	0.75	0.75
		0.7	0.2	0.12	0.08	0.37	0.05	0.08	0.10	0.07
		0.7	0.7	0.28	0.26	0.40	0.12	0.26	0.21	0.20

ρ is equal to 0.1 and 0.4. $10 \leq M_i \leq 50$

We consider a self-weighted two-stage design. We selected 5 % of PSU from each stratum with unequal inclusion probabilities proportional to the SSU sizes; that is, $\pi_i \propto M_i$. SSU are selected with equal probability with each selected PSU. One thousand two-stage samples are selected in order to calculate the empirical relative bias (%) and empirical relative root mean squared error (%) of the proposed estimator (3), the Horvitz and Thompson (1952) estimator (HT) and the Estevao and Särndal (2006) estimator defined by (1). We consider particular cases for the Estevao and Särndal (2006) estimator and the proposed estimator: (i) only the information-X is included into the estimator; (ii) only the information-Z is included into the estimator; (iii) the information-X and the information-Z are both included into the estimator.

In Tables 1 and 2, we have respectively the relative bias and the relative root mean squared error, when the PSU sizes (M_i) varies between 10 and 50. The quantity r_{yx} and r_{yz} denote the correlation between the variable of interest and the variables x and z. We clearly see that the proposed estimator is the most accurate estimator when we have a stratification by PSU size. We have more accurate estimates when the variables x and z are included into the estimators. With a random stratification, the Estevao and Särndal (2006) estimator is the most accurate. The proposed estimator is only slightly less accurate.

In Tables 3 and 4, we reduce the variability of the PSU size. In these tables, we have respectively the relative bias and the relative root mean squared error, when the PSU sizes (M_i) varies between 10 and 30. We can draw the same conclusion from these tables. However, we notice that with a random stratification the proposed estimator can be as accurate as the Estevao and Särndal (2006) estimators.

Table 2 Relative root mean square error(%) $N_1 = 3,000$, $N_2 = 2,000$, $N_3 = 1,000$, $H = 3$

		r_{yx}	r_{yz}	HT	Estevao and Särndal			Proposed		
					X	Z	XZ	XQ	ZQ	XZQ
$\rho = 0.1$	Stratification by	0.2	0.2	0.46	0.44	0.47	0.32	0.45	0.32	0.30
	PSU size	0.2	0.7	1.33	1.32	0.85	0.74	1.33	0.66	0.65
		0.7	0.2	0.15	0.11	0.41	0.07	0.11	0.12	0.06
		0.7	0.7	0.39	0.37	0.47	0.21	0.37	0.21	0.18
	Random	0.2	0.2	0.45	0.43	0.66	0.30	0.43	0.38	0.36
	stratification	0.2	0.7	1.23	1.22	0.83	0.57	1.23	0.95	0.95
		0.7	0.2	0.15	0.11	0.60	0.06	0.11	0.13	0.09
		0.7	0.7	0.34	0.32	0.61	0.15	0.32	0.26	0.25
$\rho = 0.4$	Stratification by	0.2	0.2	0.43	0.42	0.45	0.28	0.42	0.28	0.26
	PSU size	0.2	0.7	1.33	1.32	0.85	0.74	1.33	0.66	0.65
		0.7	0.2	0.15	0.10	0.41	0.06	0.10	0.12	0.06
		0.7	0.7	0.39	0.37	0.47	0.21	0.37	0.21	0.18
	Random	0.2	0.2	0.42	0.40	0.64	0.26	0.40	0.34	0.32
	stratification	0.2	0.7	1.23	1.22	0.82	0.56	1.22	0.94	0.94
		0.7	0.2	0.15	0.10	0.60	0.06	0.10	0.13	0.08
		0.7	0.7	0.34	0.32	0.61	0.15	0.32	0.26	0.25

ρ is equal to 0.1 and 0.4. $10 \leq M_i \leq 50$

Table 3 Relative bias (%) $N_1 = 3,000$, $N_2 = 2,000$, $N_3 = 1,000$, $H = 3$

		r_{yx}	r_{yz}	HT	Relative bias					
					Estevao and Särndal			Proposed		
					X	Z	XZ	XQ	ZQ	XZQ
$\rho = 0.1$	Stratification by	0.2	0.2	0.35	0.34	0.36	0.22	0.34	0.22	0.21
	PSU size	0.2	0.7	1.00	1.00	0.51	0.38	1.00	0.37	0.36
		0.7	0.2	0.11	0.08	0.31	0.04	0.08	0.09	0.04
		0.7	0.7	0.29	0.28	0.33	0.15	0.28	0.16	0.13
	Random	0.2	0.2	0.34	0.33	0.44	0.22	0.33	0.23	0.22
	stratification	0.2	0.7	0.97	0.97	0.51	0.32	0.97	0.40	0.40
		0.7	0.2	0.11	0.08	0.36	0.04	0.08	0.09	0.05
		0.7	0.7	0.26	0.24	0.37	0.07	0.34	0.12	0.10
$\rho = 0.4$	Stratification by	0.2	0.2	0.33	0.32	0.34	0.18	0.32	0.19	0.18
	PSU size	0.2	0.7	1.00	1.00	0.50	0.37	1.00	0.36	0.35
		0.7	0.2	0.11	0.08	0.30	0.04	0.08	0.09	0.04
		0.7	0.7	0.27	0.26	0.33	0.09	0.26	0.12	0.09
	Random	0.2	0.2	0.32	0.31	0.42	0.18	0.31	0.20	0.19
	stratification	0.2	0.7	0.97	0.97	0.50	0.31	0.97	0.39	0.39
		0.7	0.2	0.11	0.08	0.36	0.04	0.08	0.08	0.04
		0.7	0.7	0.26	0.24	0.37	0.07	0.24	0.12	0.10

ρ is equal to 0.1 and 0.4. $10 \leq M_i \leq 30$

Table 4 Relative root mean square error (%) $N_1 = 3,000$, $N_2 = 2,000$, $N_3 = 1,000$, $H = 3$

				Relative bias						
					Estevao and Särndal			Proposed		
		r_{yx}	r_{yz}	HT	X	Z	XZ	XQ	ZQ	XZQ
$\rho = 0.1$	Stratification by	0.2	0.2	0.44	0.43	0.51	0.28	0.43	0.28	0.27
	PSU size	0.2	0.7	1.26	1.26	0.69	0.48	1.26	0.46	0.45
		0.7	0.2	0.14	0.10	0.48	0.05	0.10	0.11	0.05
		0.7	0.7	0.37	0.35	0.47	0.19	0.35	0.20	0.17
	Random	0.2	0.2	0.43	0.41	0.63	0.28	0.41	0.29	0.28
	stratification	0.2	0.7	1.22	1.21	0.69	0.40	1.22	0.50	0.50
		0.7	0.2	0.14	0.10	0.57	0.06	0.10	0.11	0.06
		0.7	0.7	0.32	0.30	0.58	0.10	0.30	0.15	0.12
$\rho = 0.4$	Stratification by	0.2	0.2	0.42	0.41	0.49	0.23	0.41	0.24	0.22
	PSU size	0.2	0.7	1.26	1.26	0.68	0.47	1.26	0.45	0.44
		0.7	0.2	0.14	0.10	0.48	0.05	0.10	0.11	0.05
		0.7	0.7	0.34	0.33	0.51	0.12	0.33	0.15	0.11
	Random	0.2	0.2	0.41	0.39	0.62	0.23	0.39	0.25	0.23
	stratification	0.2	0.7	1.22	1.21	0.69	0.39	1.22	0.50	0.49
		0.7	0.2	0.14	0.10	0.57	0.05	0.10	0.11	0.05
		0.7	0.7	0.32	0.30	0.58	0.10	0.30	0.15	0.12

ρ is equal to 0.1 and 0.4. $10 \le M_i \le 30$

5.1 Conditional Bias

We consider to investigate the conditional performance of the proposed estimator, the Horvitz and Thompson (1952) estimator and Estevao and Särndal (2006) with the information-X & Z. We order the sampling error by their total mean and classified them into 20 groups with 50 sampled each (e.g. Chambers and Dunstan 1986). We calculate the conditional bias given $\hat{t}_{x\pi}$ or $\hat{t}_{z\pi}$. Note that in our simulation study $\hat{t}_{x\pi}$ or $\hat{t}_{z\pi}$ are scalar. We consider the following situations: the strata sizes are given by $N_1 = 3000$, $N_2 = 2000$, $N_3 = 1000$ and $r_{yx} = 0.2$, $r_{yz} = 0.7$. The results are given in Figs. 1 and 2. In Fig. 1, we have a stratification by PSU size. In Fig. 2, the PSU are allocated randomly into the strata.

In Fig. 1, we have the conditional biases when the PSU are allocated randomly into the strata. This figure shows that the proposed estimator performs well when compared to the Estevao and Särndal (2006) estimator and the Horvitz and Thompson (1952) estimators. Not surprisingly, the Horvitz and Thompson (1952) estimator has poor conditional behaviour as it shows a linear trend.

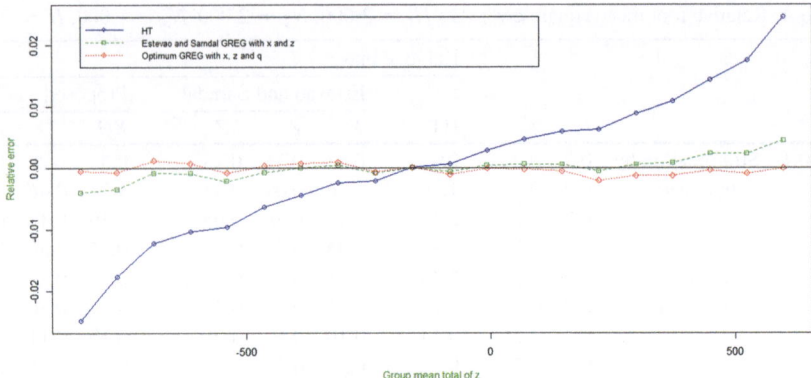

Fig. 1 Relative conditional bias for the Horvitz and Thompson (1952) estimator, the Estevao and Särndal (2006) estimator (1) and the proposed estimators (3) against the group mean total $\hat{t}_{z\pi}$. Stratification by PSU size. $r_{yx} = 0.2$ and $r_{yz} = 0.7$

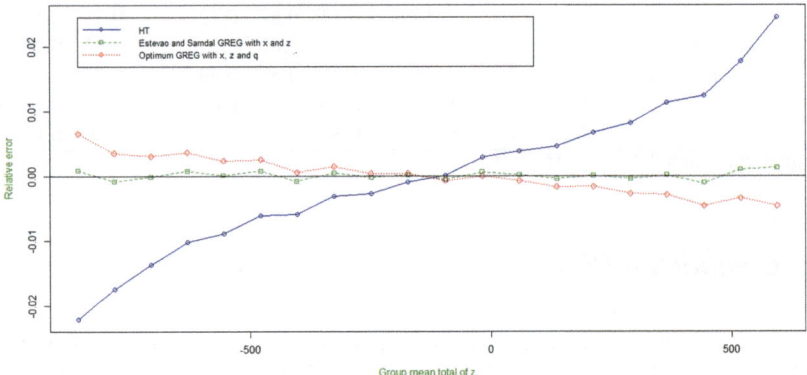

Fig. 2 Relative conditional bias for the Horvitz and Thompson (1952) estimator, the Estevao and Särndal (2006) estimator (1) and the proposed estimators (3) against the group mean total $\hat{t}_{z\pi}$. Random stratification. $r_{yx} = 0.2$ and $r_{yz} = 0.7$

In Fig. 2, we have the conditional biases for a random stratification. In this situation we see that the Estevao and Särndal (2006) estimator performs slightly better than the proposed estimator. The Horvitz and Thompson (1952) estimator also shows a linear trend.

6 Conclusions

In this paper, we propose to extend the optimal estimator proposed by Berger et al. (2003) (see also Tan 2013) for two-stage sampling. We compare the proposed estimator with the Estevao and Särndal (2006) estimator, under a self-weighted two-stage sampling design. The simulation results show that the proposed estimator may be more accurate than the Estevao and Särndal (2006) estimator when PSU of similar sizes are grouped in the same strata; that is, when the strata are homogeneous according to the PSU sizes. Note that this is a situation which is not uncommon in practice. If the strata are not related to the PSU sizes, the Estevao and Särndal (2006) estimator is slightly more accurate than the proposed estimator. In this situation, the loss of efficiency of the proposed estimator is minor.

References

Berger, Y.: Variance estimation with chao's sampling scheme. J. Stat. Plan. Inference **127**, 253–277 (2005). http://www.sciencedirect.com/science/article/pii/S0378375803002933

Berger, Y.G., Tirari, M.E.H., Tillé, Y.: Towards optimal regression estimation in sample surveys. Aust. N. Z. J. Stat. **45**, 319–329 (2003)

Chambers, R.L., Dunstan, R.: Estimating distribution functions from survey data. Biometrika **73**, 597–604 (1986)

Deville, J.: Estimation de la variance du coefficient de gini mesuré par sondage. Actes des Journées de Méthodologie Statistiques INSEE méthodes [In French] (1997)

Estevao, V.M., Särndal, C.-E.: Survey estimates by calibration on complex auxiliary information. Int. Stat. Rev. **74**, 127–147 (2006). http://dx.doi.org/10.1111/j.1751-5823.2006.tb00165.x

Hansen, M., Hurwitz, W., Madow, W.: Sample Survey Methods and Theory, Vol. I. Wiley, New York (1953)

Horvitz, D.G., Thompson, D.J.: A generalization of sampling without replacement from a finite universe. J. Am. Stat. Assoc. **47**, 663–685 (1952)

Montanari, G.: Post sampling efficient qr-prediction in large sample survey. Int. Stat. Rev. **55**, 191–202 (1987)

Narain, R.: On sampling without replacement with varying probabilities. J. Ind. Soc. Agri. Statist. **3**, 169–174 (1951)

Tan, Z.: Simple design-efficient calibration estimators for rejective and high-entropy sampling. Biometrika **100**, 399–415 (2013)

Spatial Sampling for Agricultural Data

Federica Piersimoni, Paolo Postiglione, and Roberto Benedetti

Abstract The importance of sampling spatial units is recently acknowledged in various practical studies. In most cases, spatial units are defined across a geographical domain partitioned into a number of predetermined regularly or irregularly shaped locations. In standard sampling theory, spatial units have been traditionally represented as a mosaic of areas in which individual primary units are essentially viewed as identical members of the same population. The dependence between nearest units is an inherent feature of spatial data that should be exploited at least in the sample design. In this paper, we present a review of the main topic concerning sampling spatial units. In particular, we focus our attention to spatially distributed agricultural data, which are extremely important as a support for both policy makers and market stakeholders. Our narrative aims at raising some research questions that can be explored in the near future.

Keywords Auxiliary information • Calibration methods • Small area estimation • Spatial surveys • Spatially balanced samples

1   Introduction

During the last decades we can observe an increasing agreement regarding the challenges of applying advanced survey methods to agricultural units (Benedetti et al. 2010, 2014), though in Western countries agriculture is assuming a more

F. Piersimoni (✉)
ISTAT, Agricultural Statistics Service, Rome, Italy
e-mail: piersimo@istat.it

P. Postiglione • R. Benedetti
Department of Economic Studies, University "G. d'Annunzio" of Chieti-Pescara, Italy
e-mail: postigli@unich.it; benedett@unich.it

F. Mecatti et al. (eds.), *Contributions to Sampling Statistics*, Contributions to Statistics,
DOI 10.1007/978-3-319-05320-2_12,
© Springer International Publishing Switzerland 2014

marginal economic role, if measured in terms of percentage contribution to the GDP of each country. On the other hand, agriculture is essential for the determination of livelihoods of people.

Sampling agricultural methods in agriculture can be based on a list frame or on spatial frame (FAO 1996) or on both through a multiple frame approach (Mecatti 2007). The units of a spatial frame can be points, transects (i.e., lines of a certain length) or portion of territory, often named segments. It is for this reason that a dual frame (Mecatti 2007; Lohr 2009; Singh 2014) is often considered a good solution as, if it is well designed, it may benefit from the advantages of both the frames.

The term *agricultural statistics* often includes statistics on agricultural products, forestry, fishery, livestock, and also on food safety. The definition is also related to the use of land, to the culture of a living organism through more than one life cycle, and to ownership (Everaers 2010). In agricultural statistics different research areas can be acknowledged: sample surveys, design of experiments, and biometrical techniques, among others. For a comprehensive review of statistical methods in agriculture see also Benedetti et al. (2012, 2014).

In this paper, our goal is to present the state-of-the-art methods that can be used in spatial agricultural surveys. In particular, the first aim is to review some spatial frame-based surveys that are used for agricultural data. We motivate the need for *ad-hoc* spatial techniques to achieve higher levels of efficiency. When population units are geographically distributed, classical random sampling strategies may be inefficient designs. In fact, nearby locations tend to have more similar values for measured attributes than distant ones. The choice of neighboring locations adds less additional information about the target area. So, it is clear that the definition of sampling schemes for spatial units cannot be reasonably treated unless taking into account the concept of spatial dependence.

With regard to the use of auxiliary information, Carfagna and Gallego (2005) provide a first exhaustive description of different possibilities of the use of remote sensing for agricultural statistics. In particular, remote sensing techniques may represent a suitable tool for particular problems in agricultural surveys as, for example: reliability of data, incomplete sample frame and sample size, methods of units' selection, measurement of area, non-sampling errors, gap in geographical coverage, and non-availability of statistics at disaggregated level. Remote sensing can be properly employed at the design level. Remote-sensed images provide an outline of the territory under investigation, and are useful for the construction of the spatial reference frame. Furthermore, classified satellite images can be used as auxiliary variables to improve the precision of ground survey estimates, generally with a regression or a calibration estimator. The remote-sensed information could also represent an auxiliary variable in the procedures of small area estimation (SAE).

The layout of the paper is the following. Section 2 is devoted to an overview about the use of spatial units in sampling, describing the advantages and the weakness of this sampling approach. Section 3 will contain a review of the main agricultural surveys based on spatial units. In Sect. 4 the problem of designing a spatially balanced sample, and other possible methods to optimize the design of spatial units are described. Section 5 tries to extend in a spatial frame context the use of the classical regression estimator to the more flexible calibration estimators. These model-assisted methods are then compared with small area estimators in Sect. 6 to understand the better solution to exploit remotely sensed data in agricultural surveys. Finally, Sect. 7 concludes the paper.

2 The Use of Spatial Units in Sampling of Natural and Environmental Resources

The importance of sampling spatial units is recently acknowledged in various practical problems in geographical and environmental studies. In most cases, spatial units are defined across a geographical domain partitioned into a number of predetermined regularly or irregularly shaped locations. Sample information is then used in usual operations of estimating the distinctive features of a given population when complete enumeration is too expensive.

In the first stage of a multistage sample design, the sampling frame consists of a number of large aggregate units, each of which contains subunits. We will define a first-stage unit as primary sampling unit (PSU). The investigator selects a probability sample of PSUs, and then proceeds to the second stage of sampling, in which a probability sample of subunits is selected from each PSU. These elements are defined as secondary sampling units (SSUs). Data relevant to each PSU (see the next paragraph for some examples) are frequently assumed to be independent, thus, as a consequence they are selected according to a criterion which ensures that the second-order inclusion probabilities are as close as possible to the product of the first-order probabilities. The approximation of this product is never exact since the common designs used in practice are always without replacement.

Sampling methods in agriculture can be based on a list frame or on spatial reference frame. Designs based on list frame are the most commonly used sampling procedures for agricultural surveys. The list frame is produced by an enumeration of elements of population under investigation, and in agricultural surveys is often formed by holdings or holders' addresses.

A spatial sample survey is a study in which the final-stage sampling units are land areas, and the selection probabilities are generally proportional to their area sizes. Note that, in literature, the widely used term to indicate this category of frame is area frame that can be of point or of polygon typology. In order to avoid confusion in the reader and according to the spatial statistics literature, in this paper, we prefer the general notation spatial reference or geo-coded frames, specifying in case when the frame is based on spatial points or areas.

The land areas are generally denoted as segments. The segments of a spatial frame can be areas (i.e., portion of territory), points, or transects (i.e., lines of a certain length). The sampling units should not overlap, and must cover the entire survey area under investigation.

List frame and spatial frame have advantages and drawbacks. In particular, the setup of a frame is less expensive when dealing with list frames (i.e., farms or households), since they do not require any complex GIS integration of different map layers. Furthermore, the accuracy of the survey, for a given sample size, is usually much higher for surveys based on list frames, because a farm provides in one interview more information on crop area and yields, livestock, inputs or socio-economic variables than a spatial unit. Finally, the observing/interviewing cost for each sample unit is much lower when working with list frames as in this case the time needed to fill a questionnaire is much lower than the classical cartographic operations (mapping, digitizing and so on) when observing spatial units, particularly of polygon typology.

The spatial reference frame samples (Cotter et al. 2010) are better protected against non-sampling errors due to missing or overlapping units in the frame. They do not exhibit problems linked to gaps (i.e., unbiasedness of the estimates). It is also possible to use spatial frame to verify the rate of coverage of existing archives on farms and updating them. Furthermore, the researcher can use auxiliary information (e.g., remote sensing), issue timely and higher precision estimates on cultivated areas and on expected production, and reduce burden on farmers. Finally, spatial sampling displays longevity of the frame (only updates for land use changes are necessary), versatility (multiple variables can be considered in one survey), objectivity of data collection (land cover/use and measures of areas are directly observed by surveyors in the field), low non-response rate (only for not reachable areas). The major disadvantages are represented by the high cost of setting up the frame and of sample selection and the considerable cartographic requirements to construct the frame (i.e., maps, satellite images, aerial photos). Besides, the spatial frame is not suitable for cultivations with high spatial variability (i.e., scattered), it can show limited precision of the estimates for small areas or highly concentrated land classes, and it requires well-trained enumerators and high-tech methods to qualified office staff and statisticians.

3   Agricultural Surveys Based on Spatial Reference Frames

In contrast to mapping approaches (for example, the CORINE Land Cover project), spatial sampling is a statistical method (EUROSTAT 2000). Based on the observation of sample points, the estimates are computed and used as a valid generalization without studying the entire area under investigation. Several examples of such surveys can be found in practice whose main characteristics will be shown below.

The National Agricultural Statistics Service (NASS) of the United States Department of Agriculture (USDA) has developed spatial sampling frames since 1954 as a tool for collecting information concerning crop acreage, cost of production, farm expenditures, grain yield and production, and livestock inventories (Arroway et al. 2010). The main spatial frame survey organized by the NASS is the *June Area Survey* (JAS). This mid-year survey provides area frame estimates mainly for crop acreages and livestock inventories (Nusser and House 2009). JAS includes also an estimate of the number of farms in the United States.

The information collected about specific products varies from state to state. All contributing producers provide information on their total acres, acres planted to specific commodities, and quantities of grains and oilseeds stored on farm. JAS is based on a spatial frame of areal typology. All land in the United States, except Alaska, is stratified by land use within a state. Since the disposition of crops and livestock can widely differ across a state, land is divided into homogeneous strata: intensively cultivated land, urban areas, and range land. The general strata definitions are similar in the different states. Each stratum is further divided into substrata by grouping areas that are agriculturally similar. Within each substratum, the territory is divided into PSUs. The PSUs provide complete coverage of all agriculture activities occurring within the PSU and, consequently, all farmers in the state. Each PSU is divided into segments, which are roughly a square mile in area. A sample of PSUs is selected and one segment is randomly selected from each selected PSU.

Field interviewers divide all the selected segments into tracts, where each tract represents a unique land operating arrangement. Each tract is labeled as agricultural or nonagricultural. A tract is considered as agricultural, if it has qualifying agricultural activity either inside or outside the segment. Otherwise, it is labeled as nonagricultural. Each year, about 3,500 segments are selected for inclusion in the sample. A selected segment is in the sample for 5 years.

The *Land Use/Cover Area frame Survey* (LUCAS) (Gallego and Delincè 2010) is a project funded by EUROSTAT that was initially introduced to offer, on a yearly basis, European crop estimates. Across time, this survey has also provided data on land use, being a valuable tool for environmental monitoring. The main objectives of the LUCAS project are to obtain harmonized data (in particular unbiased estimates) at EU country level of the main land use and cover areas and their changes; to increase the scope of the survey, beyond the usual agricultural domain; to include also aspects related to environment, landscape, and sustainable development; to provide a common sampling base (i.e., frame, nomenclature, data treatment) that interested Member States can use to obtain representative data at national/regional level. The sample frame is performed at country level, because there is no possibility to create a regular grid over the complete territory of Europe for statistical purposes. The sample frame is based on official digital geographic data of administrative boundaries and coastlines of Europe available at EUROSTAT GISCO (the *Geographical Information System at the COmmission*).

The LUCAS is a spatial reference frame survey based on points sampling, and it was carried out in 2001 and 2003 for EU15 based on a systematic sample of almost 100,000 points grouped in clusters of ten points. The LUCAS has a double nomenclature: each point has a land cover code (57 classes) and a land use code (14 classes). In 2006, the design was modified in a stratified two-phase sampling of unclustered points (Jacques and Gallego 2005). The stratification was performed by photo-interpretation of a 2-km grid of points in EU25. The points placed on small islands were not considered in the sample (i.e., Baleares, Azores, Canary

Islands, Cyprus, Malta, and the Greek Islands except Crete). The sampling design has been again modified for the LUCAS 2009 for giving more importance to the environmental and agri-environmental parameters. The choice of maximizing the distance of the points, both in the same and in different strata, was outlined according to the 2006 wave, since it performed to be efficient (Jacques and Gallego 2005). On the other hand, points sampled in different strata can be close to each other and give some extra information concerning the presence of spatial correlation between strata. To reduce the effect of autocorrelation within and between strata, the basic sampling grid (2×2 km^2) has been divided into squared (9×9 that means 18 km by 18 km each) blocks of 81 points each. The set of points with the same relative position in the block is named as a replicate. The numbering of the replicates is done under the constraint that the distance with the previous ones is maximized. Replicates are then chosen successively until the required sample size by domain is obtained. From the replicate with the highest number, points are randomly selected. The above described selection method was combined with a panel approach. Finally, points with an altitude above 1,000 m were excluded from the second phase sample, since it is assumed that their importance in agriculture is very poor. Currently, EUROSTAT has carried out the LUCAS 2012 survey in the European Union. The LUCAS 2012 covers all 27 EU countries. The fieldwork has been carried out in March–September 2012.

The AGRIT Italian program (Carfagna and Gallego 2005; Postiglione et al. 2010) is a spatial sampling frame survey that aims at providing conjunctural estimates on areas, on yields of the main crops, and on the main land uses. This survey is realized through techniques of spatial sampling, especially of point typology. The list is formed by a set of points, each point is provided of an operational dimension area: a 3 m radius circle centered on the point (for a total of about 30 m^2). The sampling design is stratified in two or three phases. The sample of the first phase comprises an aligned spatial systematic selection of geographical units. This list of units is indicated as an AGRIT sampling frame, and is formed by a regular grid of points (about 1,200,000), selected with a step of 500 m defined in the coordinate system Gauss-Boaga for a complete coverage of the land under investigation. The sampling units of second phase are based on a preliminary identification of zones of interest for the analysis at the different levels of administrative division: province, region, and the entire national territory. Finally, the yield estimates of the third sampling phase are significant with reference to the same geographical levels above described, but for only 12 different crops. An archive of land uses available for agricultural surveys may derive from three different sources: ground survey, interpretation of digital orto-photos, and the classification of remotely sensed images. This activity, which formed the hierarchical nomenclature of reference (on two levels for a total of 25 labels), was developed within the *Permanent Observed POints for Land Use Statistics* (POPOLUS) project. The 25 POPOLUS classes were aggregated into six groups, only four then sampled.

The first step in the preparation of the archive was the stratification of the PSUs. The points were stratified through the Italian 103 provincial codes and 6 classes of land use, by obtaining 618 not empty strata. The strata codes for land uses are: arable land, permanent crops, permanent fodder land (altitude $\leq 1,200$ m), permanent fodder land (altitude $>1,200$ m), wooded land, isolated trees and agricultural buildings, and other (i.e., artificial surfaces, water, not vegetated natural surfaces). All the points that display an agricultural activity compose the target population for the selection of the second phase sample. The total size consists of approximately 80,000–150,000 points to be collected on the ground depending on the year of the survey. The survey has remained essentially unchanged across the years.

The knowledge and the monitoring of the land use and cover are old issues in France. The *Utilisation du territoire* (TER-UTI) survey is annually carried out by the statistical services of the French Ministry of Agriculture and Fishing to collect data on land use of the whole continental territory by using a set of points constituting a representative sample of the territory. The first TER-UTI was carried out in 1982. It was renewed in 1990 and 1991 to correct certain bias introduced during sample selection in 1982. In 2005 the survey design was re-defined. Two main reasons have led the researchers to modify the survey. First, technical enhancement in digitization of cartographic and geo-referencing points motivated re-defining the sampling points TER-UTI, which were previously determined manually on aerial photographs. Second, the introduction of the LUCAS survey by EUROSTAT, based on the same methodological principles as TER-UTI, has led to a deep revision that allows for consistency of nomenclature, the method of observation, and the design of the sample. For these reasons, the French survey is now called as TER-UTI LUCAS. It uses a non-stratified two-stage sampling scheme, with points grouped in PSUs. The PSUs are the segments, generally identified by a square area that varies from 1.5 km by 600 m to 1.5 km by 1.5 km. The second-stage units are the points, with 3 m radius circle centered on the point (basic observation window) in the general case or with a 40 m radius circle (extended observation window) in the case of heterogeneous land cover. Points are spaced 300 m inside a segment.

4   Methods for Using Remote Sensing Data at the Design Level

Surveys are routinely used to gather primary data in agricultural research. The units to be observed are often randomly selected from a finite population whose main feature is to be geo-referenced. Thus, its spatial distribution has been widely used as crucial information in designing the sample.

In business surveys in general, and in multipurpose agricultural surveys in particular, the problem of designing a sample from a frame usually consists of three different aspects. The first is concerned with the choice of a rule for stratifying the population when several size variables are available. The second refers to the definition of the selection probabilities for each unit in the frame. Finally, the third is devoted to sample size determination and sample allocation to a given set of strata. The main required property of the sample design is that it should provide a specified level of precision for a set of variables of interest using as few sampling units as possible. Stratification is introduced into sampling designs for a number of different reasons: for example, to select the sample from a given frame, to obtain reliable estimates for subpopulations (i.e., domains), or to improve the estimators' efficiency of global population parameters. In most cases, populations are either naturally stratified or can be easily stratified on the basis of practical considerations such as administrative subdivisions. In other situations, strata are established in order to satisfy interest in identifying characteristics of subpopulations. When such straightforward definitions are not possible, then a decision ought to be taken on the number of strata and their respective boundaries.

These classical design issues should be considered together with the importance of selecting samples of statistical units taking into account their geographical position. This issue is now more than ever recognized in the measuring process of several phenomena for various reasons. First, because of the evidence that the statistical units are defined by using purely spatial criteria, as in most agricultural and environmental studies. Second, in many countries there is a common practice that the National Statistical Institute (NSI) geo-references the typical sampling frames of physical or administrative bodies not only according to the codes of a geographical nomenclature but also adding information regarding the exact, or estimated, position of each record. Even if in natural resources monitoring and estimation infinite populations cover an important part of the sampling problems, in agricultural surveys we mainly deal with finite populations. In this context, the spatial distribution of the frame is a strong constraint, and for this reason we suspect that it could have a considerable impact on the performance of a random sampling method. For example, the traditional solution of extending the systematic sampling to multidimensional data by simply overlaying a grid of points to a spatial domain could not be feasible, if the population is far to be considered distributed on a regular grid as it is clustered or it shows to have different intensities of the units across the domain.

Assume that we are interested in estimating some parameter of a set $\mathbf{Y} = \{\mathbf{y}_1, \ldots, \mathbf{y}_j, \ldots, \mathbf{y}_v\}$ of v target variables where $\mathbf{y}_j$ is the generic j-th variable, and let $U = \{1, 2, \ldots, N\}$ be a finite population recorded on a frame together with a set of m auxiliary variables $\mathbf{X} = \{\mathbf{x}_1, \ldots, \mathbf{x}_j, \ldots, \mathbf{x}_m\}$ and a set of h (usually $h = 2$) coordinates $\mathbf{C} = \{\mathbf{c}_1, \ldots, \mathbf{c}_j, \ldots, \mathbf{c}_h\}$ obtained by the geo-coding of each unit, where $\mathbf{x}_j = \{x_{1j}, \ldots, x_{kj}, \ldots, x_{Nj}\}$ is the generic j-th auxiliary and $\mathbf{c}_j = \{c_{1j}, \ldots, c_{kj}, \ldots, c_{Nj}\}$ is the generic j-th coordinate. From $\mathbf{C}$ we can always derive, according to any distance definition, a matrix $\mathbf{D}_U = \{d_{kr}; k = 1, \ldots, N, r = 1, \ldots, N\}$, which specifies how far are all the pairs of units in the population.

The geographical position in many agricultural surveys is an intrinsic characteristic of the unit, and its efficient use in sample design often requires methods that cannot be adapted from those used when dealing with classical auxiliary variables. This is not only a consequence of its multivariate nature and the traditional design solutions, as the πps (i.e., inclusion probability proportional to size), can handle only one auxiliary (Bee et al. 2010). To use some covariates, we always assume that there is, at least approximate, a certain degree of correlation between a survey variable **y** and the set **X**. While, with regard to the use of the set **C**, the distance matrix, as a synthesis of the spatial information, emphasizes the importance of the spread of the sample over the study region. This feature can be related, but not necessarily, to the dependence but also to some form of similarity between adjacent units.

Usually **X** and **C** in agricultural surveys play different roles according to the definition of the statistical unit:

1. When U is a list of agricultural households, **C** is rarely obtainable, depending on the availability of accurate cadastral maps, and should be constituted by a map of polygons representing parcels of land used by each holding. **X** is usually filled with administrative data sources, previous census data and, only if **C** is available, remotely sensed data obtained through the overlay of the polygon map with a classified image.
2. If U is a list of regularly or irregularly shaped polygons defined *ad hoc* for the agricultural survey, **C** is always available since it represents the definition of each statistical unit and **X**, unless an overlay of **C** with a cadaster is possible, can be constituted only by some geographical coding and summarizing a classification arising from remotely sensed data within each polygon.
3. Another possible choice, widely used in agricultural surveys, is that U is a list of points, usually the corners of regular grid overlaid over the survey geographical domain, which, thus, does not represent an exhaustive population of the study area but only the first stage of sampling. In this case, **X** can be only represented by a geographical nomenclature and by a design matrix of codes of land use classification obtained, or by previous land use maps, or by a classification of remotely sensed data while **C** is simply the coordinates of each point.

In surveys of type 1., the relevant characteristic to be controlled is that the target population is very skewed as the concentration of farms and agricultural households sizes is very high. In sampling theory, the large concentration of the population with respect to surveyed variables constitutes a problem that is difficult to handle without the use of selection probabilities proportional to a size measure or by the use of a stratification or partition tool.

4.1   Multivariate Auxiliaries in πps Sampling

One of the methods for the utilization of auxiliary information *ex ante* is to employ a sampling scheme with inclusion probabilities proportional to given size measures, a so-called πps scheme (Rosén 1997). This sampling scheme has desirable properties, but it cannot be applied in practical situations where the frame contains a multivariate **X**, because it is seriously limited by the drawback that the method can use only one auxiliary variable (Benedetti et al. 2010, 2014).

The design of a πps random sample from a finite population, when multivariate auxiliary variables are available, deals with two main issues: the definition of a selection probability for each unit in the population as a function of the whole set of the auxiliary variables, and the determination of the sample size required to achieve a constrained precision level for each auxiliary variable. These precisions are usually expressed as a set of upper limits on the coefficients of variation of the estimates. Deville and Tillé (1998) suggested some interesting solutions to the problem of selecting a sample by using a πps scheme. Chauvet and Tillé (2006) reviewed the application of several πps algorithms. However, their focus was mainly on how to respect the defined probabilities and the performance of each selection procedure. These classical methods work with the univariate case and cannot be easily extended to treat the case, often observed in real circumstances, particularly in agricultural surveys, where it is important to deal with a multipurpose survey and to exploit in the sampling design multiple covariates such as land use classes arising from a remotely sensed data classification.

We suggest a solution for a πps scheme that can consider more auxiliaries in the sample selection process. We refer to this approach as a multivariate πps. As stated above, a general methodological framework to handle this situation is missing in past literature, even if several practical efforts have been already made in this direction: in some NASS–USDA's surveys the use of the maximum probability was suggested (Bee et al. 2010). In a previous work some good results in terms of root mean square error (RMSE) have been obtained by simply defining the vectors of first-order inclusion probabilities as the averages of such probabilities for each auxiliary variable (Bee et al. 2010).

Moreover, it is important to mention that there are other ways to consider multiple auxiliary variables in the sample selection procedure (Bee et al. 2010). In particular, the Cube method for balanced sampling (Chauvet and Tillé 2006; Deville and Tillé 2004; Tillé 2011; Tillé and Favre 2005), with constant or varying inclusion probabilities, can be used to select a sample that satisfies a given vector of selection probabilities and that is, at the same time, balanced on a set of auxiliary variables. Following these considerations, some recent studies were focused on the computation of optimal inclusion probabilities for balanced sampling on given auxiliary variables (Tillé and Favre 2005).

4.2 Optimal Stratification

A traditional approach to take into account multivariate auxiliary variables in designing the sample is to employ a stratification scheme such that the population units are classified in a stratum according to the values of their auxiliary variables (Benedetti et al. 2008; Vogel 1995). Most of the literature on *optimal stratification* relies on the early works of Dalenius and Hodges (see Horgan 2006 for a review) in the 1950s, whose solutions, usually based on linear programming, are still widely popular in applied survey sampling (Khan et al. 2008).

In the literature several formal extensions to the univariate optimal determination of the boundaries between more than two strata have been proposed (Kozak 2004; Horgan 2006), through the use of algorithms that usually derive simultaneously the sample size needed to guarantee a fixed accuracy level for the resulting estimates and the sample allocation to the strata. A generalization of these algorithms, extended in Baillargeon and Rivest (2011), is used when the survey variable and the stratification variable differ. These classical methods deal only with the univariate case, and cannot be easily extended when there are multiple covariates for stratification. Within a multivariate context, the use of stratification trees (Benedetti et al. 2008) has several advantages over that of classical univariate Dalenius-type methods. First, stratification trees do not require either distributional assumptions about the target variable, or any hypotheses regarding the functional form of the relation between this variable and the covariates. Moreover, when many auxiliary variables are available, the stratification tree algorithm is able to automatically select the most powerful variables for the construction of strata. The identified strata are easier to interpret than those based on linear methods. Finally, they do not require any particular sample allocations to the strata, since it simultaneously allocates the sampling units using the Bethel or the Cromy algorithm in each iteration (Benedetti et al. 2008).

However, such an approach is equivalent to partitioning the population into strata that have *box-shaped* boundaries or that are approximated through the union of several such boxes. This constraint prevents the identification of irregularly shaped strata boundaries unless a grid constituted by several rectangles of different size is used to approximate the required solution.

Optimal data partitioning is a classical problem in statistical literature. Note that our problem is more directly related to the use of unsupervised classification methods to cluster a set of units (i.e., a population frame). The main difference between the two problems lies in the fact that the underlying objective functions are different. In sampling design the aim is usually to minimize the sample size. On the other hand, in clustering it is a common practice to minimize the within cluster variance. There is an intuitive connection between these two concepts even if the definition of sample size depends not only on the variance within each stratum, but also on other parameters (for example, sample size, population size, unknown total).

4.3   Spatially Balanced Samples

In the last decades, the spatial balancing of samples has become so important that several sampling algorithms have been introduced to realize this objective by researchers and survey practitioners (Wang et al. 2012).

In design-based sampling theory, if we assume that there is no measurement error, the potential observations over each unit of the population cannot be considered dependent. However, an inherent and fully recognized feature of spatial data is that of being dependent as shortly expressed in Tobler's *first law of geography*, according to which *everything is related to everything else, but near things are more related than distant things*. It is then clear that sampling schemes for spatial units can be reasonably treated by introducing a suitable model of spatial dependence within a model-based or at least a model-assisted framework. In past literature (Benedetti and Palma 1995; Rogerson and Delmelle 2004), this approach proved to be helpful to find a rationale for the intuitive procedure to spread the selected units over the space, because closer observations will provide overlapping information as an immediate consequence of the dependence. Under this assumption the concern is necessarily in finding the sample configuration, which is the best representative of the whole population, and leads to define our selection as a combinatorial optimization problem. In fact, provided that the sample size is fixed, the aim is that of minimizing an objective function defined over the whole set of possible samples, which represents a measure of the loss of information due to dependence. An optimal sample selected with certainty is of course not acceptable if we assume the randomization hypothesis, which is the background for design-based inference. Thus, we should move from the concept of dependence to that of spatial homogeneity measured in term of local variance of the observable variable, where for local units we could define all the units of the population within a given distance. An intuitive way to produce samples that are well spread over the population is to stratify the units of the population on the basis of their location. The problems arising by adopting this strategy lie in the evidence that it does not have a direct and substantial impact on the second-order inclusion probabilities, surely not within a given stratum, and that frequently it is not clear how to obtain a good partition of the study area. These drawbacks are in some way related and for this reason they are usually approached together by defining a maximal stratification, (i.e., partitioning the study in as many strata as possible and selecting one or two units per stratum). This simple and quick scheme, to guarantee that the sample is well-spread over the population, is somewhat arbitrary, because it highly depends on the stratification criterion that should be general and efficient.

The basic principle is to extend the use of systematic sampling to two or more dimensions, an idea that is behind the generalized random tessellation stratified (GRTS) design (Stevens and Olsen 2004) that, to systematically select the units, maps the two-dimensional population into one dimension while trying to preserve some multidimensional order.

This approach is essentially based on the use of Voronoi polygons that are used to define an index of *spatial balance*. For a generic sample s the Voronoi polygon for the sample unit $s_k = 1$ includes all population units closer to s_k than to any other sample unit $s_r = 1$. If we let v_k be the sum of the inclusion probabilities of all units in the k-th Voronoi polygon, for any sample unit u_k, we have $E(v_k) = 1$ and for a spatially balanced sample all the v_k should be close to 1. Thus, the index $V(v_k)$ (i.e., the variance of the v_k) can be used as a measure of spatial balance for a sample.

In a spatial context, the balanced sampling (Deville and Tillé 2004), through the introduction of the cube method (Chauvet and Tillé 2006), could be applied by imposing that any selected sample should respect for each coordinate the first p moments, assuming implicitly that the survey variable y follows a polynomial spatial trend of order p (Breidt and Chauvet 2012). In the last paper there is also an interesting application to spatial sampling using kriging. Besides, Deville and Tillé (2004) and Chauvet and Tillé (2006) do not use the concept of distance, which is a basic tool to describe the spatial distribution of the sample units, which leads to the intuitive criterion, that units that are close, seldom appear simultaneously in the sample.

Following this line Arbia (1993), inspired by purely model-based assumptions on the dependence of the stochastic process generating the data, according to the algorithm typologies identified by Tillé (2006), suggested a draw-by-draw scheme, the dependent areal units sequential technique (DUST), which starting with a unit selected at random, say k, in any step $t < n$ updates the selection probabilities according to the rule $\pi_r^{(t)} = \pi_r^{(t-1)}\left(1 - e^{-\lambda d_{kr}}\right)$, where λ is a tuning parameter useful to control the distribution of the sample over the study region. This algorithm can be easily interpreted and analyzed in a design-based perspective in particular referring to a careful estimation and analysis of its first- and second-order inclusion probabilities.

Very recently, some advances have been proposed for list sequential algorithms whose updating rules have the crucial property to preserve the fixed first-order inclusion probabilities (Grafström 2012; Grafström et al. 2012; Grafström and Tillé 2013). In particular, Grafström (2012) suggested a list sequential algorithm that, for any unit i, in any step t, updates the inclusion probabilities according to a rule $\pi_k^{(t)} = \pi_k^{(t-1)} - w_t^{(k)}(I_t - \pi_t^{(t-1)})$, where $w_t^{(k)}$s are weights given by unit t to the units $k = t + 1, t + 2, \ldots, N$ and I_t is an indicator function set equal to 1 if the unit t is included in sample and equal to 0 otherwise. The weight determines how the inclusion probabilities for the unit k should be affected by the outcome of unit t. They are defined in such a way that, if they satisfy an upper and a lower bound, the initial π_k is not modified. The suggested *maximal weights* criterion gives as much weight as possible to the closest unit, then to the second closest unit and so on.

Two alternative procedures to select samples with fixed π_k and correlated inclusion probabilities were derived (Grafström et al. 2012) as an extension of the Pivotal method introduced to select πps (Deville and Tillé 1998). They are essentially based on an updating rule of the probabilities π_k and π_r that at each step should locally keep the sum of the updated probabilities as constant as possible and differ from each other in a way to choose the two nearby units k and r. These two methods are referred to as the Local Pivotal Method 1 (LPM1), which, according to the authors' suggestion is the more balanced of the two, and the Local Pivotal Method 2 (LPM2), which is simpler and faster.

There could be a lot of reasons why it will be appropriate to put some effort on selecting samples, which are spatially well distributed:

1. y has a linear or monotone spatial trend.
2. There is spatial autocorrelation (i.e., close units have more similar data than distant units).
3. y shows to follow zones of local stationarity of the mean and/or of the variance, or in other words, a spatial stratification exists in observed phenomenon.
4. The units of the population have a spatial pattern, which can be clustered, or in other words, the intensity of the units varies across the study region.

It is worth noticing that, while the distance between a couple is a basic concept in all these features of the phenomenon, the index $V(v_k)$ of spatial balance seems to be directly related to the third aspect but only indirectly to the other three. This consideration and the practical impossibility to use the index $V(v_k)$, as it involves the π_k, suggests the use of a rule that sets the probability $p(s)$ to select a sample s proportionally, or more than proportionally, to a synthesis of the distance matrix within the sample d_s.

5 Use of the Calibration Estimators

Apart from the difficulties typical of business data, such as the quantitative nature of many variables and their high concentration, agricultural surveys are characterized by some additional peculiarities. In the case of auxiliary information, there are two specific issues that need to be discussed. First, the definition of the statistical units is not unique. The list of possible statistical units is quite large, and its choice depends not only on the phenomenon for which we are interested in collecting the data, but also on the availability of a frame of units. Second, a rich set of auxiliary variables, other than dimensional variables, is available: consider, for example, the information provided by airplane or satellite remote sensing.

Concerning the first issue, agricultural surveys can be conducted using list frame or spatial reference frame. A list frame is generally based on the agricultural census, farm register, or administrative data. A spatial reference frame is defined by a cartographic representation of the territory and by a rule that defines how it is divided into units. According to the available frame, we can have different statistical units. With regard to the second issue, the rich set of auxiliary variables in agricultural surveys is mainly available from remotely sensed data. Remote sensing can significantly contribute to provide a timely and accurate picture of the agricultural sector, as it is very suitable for gathering information over large areas with high revisit frequency. Indeed, a large range of satellite sensors regularly provides data covering a wide spectral range. For deriving the sought information, a large number of spectral analysis tools have been developed.

A commonly used auxiliary variable for crop area estimates is the land use/land cover (LULC) data. LULC refers to data that are a result of raw satellite data classification into categories based on the return value of the satellite image. LULC data are most commonly in a raster or grid data structure, with each cell having a value that corresponds to a certain classification. LULC have been widely applied to estimate crop area. Hung and Fuller (1987) combined data collected by satellite with data collected with area surveys to estimate crop areas. Basic survey regression estimation is compared with two methods of transforming the satellite information prior to regression estimation. Remote sensing data provide also information on different factors that influence the crop yield. The most popular indicator for studying vegetation health and crop production is the NDVI, which is a normalized arithmetic combination of vegetation reflectance in the red and near-infrared.

The availability of remote sensing data does not eliminate the need for ground data, since satellite data do not always have the accuracy required. However, this information can be used as auxiliary data to improve the precision of the direct estimates. In this framework, the calibration estimator can improve the efficiency of crop areas and yield estimates for a large geographical area, when classified satellite images and NDVI can be used as auxiliary information respectively.

The technique of estimation by calibration was introduced by Deville and Särndal (1992). The idea behind is to use auxiliary information to obtain new sampling weights, called calibration weights that make the estimates coherent with known totals. The estimates are generally design consistent and with smaller variance than the HT estimator. The link between the variables of interest and the auxiliary information is very important for a successful use of the auxiliary information. In agricultural surveys, there are differences among the statistical units regarding the use of the auxiliary variables available.

When the statistical units are the agricultural holdings, the use of the auxiliary information depends on the availability of their positioning. If the agricultural holdings are geo-referenced the vector of auxiliary information for crop area estimates related to the farm k is given by $\mathbf{x}_k$ with containing the number of pixels classified in crop type j according to the satellite data in the farm k. When the statistical units are points, the vector of auxiliary information for crop area estimates related to the point k is given by $\boldsymbol{\delta}_k = (\delta_{k1}, \ldots, \delta_{kj}, \ldots, \delta_{kJ})^t \ k = 1, \ldots, N$, where $\delta_{(kj)}$ is a generic indicator variable with value $\delta_{(kj)} = 1$ if the point k is classified in crop type j and $\delta_{(kj)} = 0$ otherwise.

The location accuracy between the ground survey and satellite images and the difficulties in improving this accuracy through geometrical correction have been considered one of the main problems in relating remote sensing satellite data to crop areas or yields, mainly in point frame sample surveys where the sampled point represents a very small portion of the territory.

When the statistical units are regular or irregular polygons, similarly to agricultural holdings, the vector of auxiliary information for crop area related to the point k is given by the number of pixels classified in crop type j according to the satellite data in the point k. The calibration weights are justified primarily by their consistency with the auxiliary variables. However, statisticians are educated to think in terms of models, and they feel obligated to always have a statistical procedure that states the associated relationship of y to $\mathbf{x}$. This consideration leads to the idea of model calibration as it is proposed in Wu and Sitter (2001), Wu (2003), and Montanari and Ranalli (2005).

6   Small Area Estimators

The term small area (SA) generally refers to a small geographical area or a spatial population unit for which reliable statistics of interest cannot be produced due to certain limitations of the available data. For instance, SAs include small geographical regions like county, municipality, or administrative division where the domain-specific sample size is considered small. SAE is a research topic of great importance, because of the growing demand for reliable small area statistics even when only very small samples are available for these areas. The problem of SAE is twofold. The first issue is represented by how to produce reliable estimates of characteristics of interest for small areas or domains, based on very small samples taken from these areas. The second issue is how to assess the estimation error of these estimates.

In the context of agriculture, the term small area usually refers to crop areas and crop yields estimates at small geographical area levels. Agriculture statistics are generally obtained through sample surveys where the sample sizes are chosen to provide reliable estimators for large areas. A limitation of the available data in the target small areas seriously affects the precision of estimates obtained from area-specific direct estimators.

When auxiliary information is available, the design-based regression estimator is a classical technique used to improve the precision of a direct estimator. This technique has been widely applied to improve the efficiency of crop area estimates, where the used auxiliary information is given by satellite image data. Unfortunately, direct area-specific estimates may not provide acceptable precision at the SA level, in other terms they are expected to return undesirable large standard errors due to the small size (or even zero) of the sample in the area. Furthermore, when there are no sample observations in some of the relevant small domains, the direct estimators cannot even be calculated. In order to increase precision of area-specific direct estimators, various types of estimators that combine both the survey data for the target small areas and auxiliary information from sources outside the survey, such as data from a recent census of agriculture, remote sensing satellite data and administrative records, have been developed. Such estimators, referred as indirect estimators, are based on models (implicit or explicit) that provide a link to related small areas through auxiliary data, in order to borrow information from the related small areas and thus increase the effective sample size. Many contributions have been introduced in literature on the topic of SAE. In particular, the paper of Ghosh and Rao (1994), Rao (2003), and Pfeffermann (2013) have highlighted the main theories on which the practical use of small area estimator is based on.

Indirect small area estimates that make use of explicit models for taking into account specific variation between different areas have received a lot of attention for several reasons:

1. The explicit models used are a special case of the linear mixed model and thus are very flexible for handling complex problems in SAE (Fay and Herriot 1979; Battese et al. 1988).
2. Models can be validated from sample data.
3. The MSE of the prediction is defined and estimated with respect to the model.

Small area estimates that make use of explicit models are generally referred as Small Area Models (SAM) and they can be broadly classified into two types: area-level models and unit-level models (Rao 2003).

Area-level models such as the Fay and Herriot (1979) model are widely used to obtain efficient model-based estimators for small areas in agriculture statistics where a rich set of auxiliary variables is mainly available from remote sensing data. Benedetti and Filipponi (2010) addressed the problem of improving the land cover estimate at a small-area level related to the quality of the auxiliary information. Two different aspects associated with the quality of remote sensing satellite data have been considered:

1. The location accuracy between the ground survey and satellite images.
2. Outliers and missing data in the satellite information.

The first problem is addressed by using the area-level model. The small-area direct estimator is related to area-specific auxiliary variables, that is, number of pixels classified in each crop type according to the satellite data in each small area. The missing data problem is addressed by using a multiple imputation.

Furthermore, various extensions have been proposed. Datta et al. (1991) proposed the multivariate version of the Fay and Herriot model that led to more efficient estimators. Rao and Yu (1994) suggested an extension of area-level models for the analysis of time-series and cross-sectional data. Pratesi and Salvati (2008) introduced the spatial dependencies within the area-level models, and Benedetti et al. (2013) assumed that a hidden partition of the study region would lead to a local stationarity of the model parameters.

For further details on recent developments about SAE, the reader can see the special issue recently published by the Journal of the Indian Society of Agricultural Statistics (AA. VV. 2012).

7   Conclusions

Spatial sampling has a long tradition, but only recently it has been reconsidered in the view of the various problems arising in spatial data analysis. Pointing out the significance of the inherent structure of dependence, which characterizes spatial data, several authors in latest contributions (Stevens and Olsen 2004; Grafström 2012; Grafström et al. 2012; Grafström and Tillé 2013) claim that spatial units cannot be sampled as if they were generated within the classical urn model. Moreover, the parallel growth of the modern technology of remote sensing and GIS has brought into evidence the possibility of dealing with large quantities of geographical data which can be easily accessed and manipulated and thus used as auxiliary variables in the design of a sample and in the estimation process.

This extensive review aims at describing some possible research lines that can be analyzed by survey statisticians in the near future. In particular, we think that, among others, some topics require great attention and research efforts. For example, it is necessary to analyze with more details some issue at design level as optimal stratification, multivariate πps, and spatial sampling methods. Finally, it is very interesting to investigate some nonlinear extensions to calibration and small area estimation methods.

References

AA. VV.: Special issue on small area estimation. J. Indian Soc. Agric. Stat. **66**, 1–239 (2012)

Arbia, G.: The use of GIS in spatial statistical surveys. Int. Stat. Rev. **61**, 339–359 (1993)

Arroway, P., Abreu, D.A., Lamas, A.C., Lopiano, K.K., Young, L.Y.: An alternate approach to assessing misclassification in JAS. In: Proceedings of the Section on Survey Research Methods JSM 2010, American Statistical Association, Alexandria (2010)

Baillargeon, S., Rivest, L.-P.: The construction of stratified designs in R with the package stratification. Surv. Methodol. **37**, 53–65 (2011)

Battese, G.E., Harter, R.M., Fuller, W.A.: An error component model for prediction of county crop areas using survey and satellite data. J. Am. Stat. Assoc. **83**, 28–36 (1988)

Bee, M., Benedetti, R., Espa, G., Piersimoni, F.: On the use of auxiliary variables in agricultural surveys design. In: Benedetti, R., Bee, M., Espa, G., Piersimoni, F. (eds.) Agricultural Survey Methods, pp. 107–132. Wiley, Chichester (2010)

Benedetti, R., Filipponi, D.: Estimation of land cover parameters when some covariates are missing. In: Benedetti, R., Bee, M., Espa, G., Piersimoni, F. (eds.) Agricultural Survey Methods, pp. 213–230. Wiley, Chichester (2010)

Benedetti, R., Palma, D.: Optimal sampling designs for dependent spatial units. Environmetrics **6**, 101–114 (1995)

Benedetti, R., Espa, G., Lafratta, G.: A tree-based approach to forming strata in multipurpose business surveys. Surv. Methodol. **34**, 195–203 (2008)

Benedetti, R., Bee, M., Espa, G., Piersimoni, F.: Agricultural Survey Methods. Wiley, Chichester (2010)

Benedetti, R., Piersimoni, F., Postiglione, P.: Statistics in agricultural production. In: Balakrishnan, N. (ed.) Methods and Applications of Statistics in the Social and Behavioral Sciences, pp. 357–373. Wiley, Hoboken (2012)

Benedetti, R., Pratesi, M., Salvati, N.: Local stationarity in small area estimation models. Stat. Methods Appl. **22**, 81–95 (2013)

Benedetti, R., Piersimoni, F., Postiglione, P.: Sampling Spatial Units for Agricultural Surveys. Springer (2014, forthcoming)

Breidt, F.J., Chauvet, G.: Penalized balanced sampling. Biometrika **99**, 945–958 (2012)

Carfagna, E., Gallego, F.J.: Using remote sensing for agricultural statistics. Int. Stat. Rev. **73**, 389–404 (2005)

Chauvet, G., Tillé, Y.: A fast algorithm of balanced sampling. Comput. Stat. **21**, 53–62 (2006)

Cotter, J., Davies, C., Nealon, J., Roberts, R.: Area frame design for agricultural surveys. In: Benedetti, R., Bee, M., Espa, G., Piersimoni, F. (eds.) Agricultural Survey Methods, pp. 169–192. Wiley, Chichester (2010)

Datta, G.S., Fay, R.E., Ghosh, M.: Hierarchical and empirical Bayes multivariate analysis in small area estimation. In: Proceedings of Bureau of the Census 1991 Annual Research Conference, pp. 63–79. US Bureau of the Census, Washington (1991)

Deville, J.C., Särndal, C.-E.: Calibration estimators in survey sampling. J. Am. Stat. Assoc. **87**, 376–382 (1992)

Deville, J.C., Tillé, Y.: Unequal probability sampling without replacement through a splitting method. Biometrika **85**, 89–101 (1998)

Deville, J.C., Tillé, Y.: Efficient balanced sampling: The cube method. Biometrika **91**, 893–912 (2004)

EUROSTAT: Manual of concepts on land cover and land use information systems. In: Theme 5: Agriculture and Fisheries: Methods and Nomenclatures. Office for Official Publications of the European Communities, Luxembourg (2000)

Everaers, P.: The present state of agricultural statistics in developed countries: Situation and challenges. In: Benedetti, R., Bee, M., Espa, G., Piersimoni, F. (eds.) Agricultural Survey Methods, pp. 1–24. Wiley, Chichester (2010)

FAO: Multiple frame agricultural surveys. v. 1: Current surveys based on area and list sampling methods. Statistical Development Series n. 7. http://www.fao.org/fileadmin/templates/ess/ess_test_folder/Publications/SDS/sds_7_Multiple_frame_agricultural_surveys_01.pdf (1996)

Fay, R.E., Herriot, R.A.: Estimates of income for small places: An application of James–Stein procedures to census data. J. Am. Stat. Assoc. **74**, 269–277 (1979)

Gallego, J., Delincè, J.: The European land use and cover area-frame statistical survey. In: Benedetti, R., Bee, M., Espa, G., Piersimoni, F. (eds.) Agricultural Survey Methods, pp. 151–168. Wiley, Chichester (2010)

Ghosh, M., Rao, J.N.K.: Small area estimation: an appraisal. Stat. Sci. **9**, 55–93 (1994)

Grafström, A.: Spatially correlated Poisson sampling. J. Stat. Plan. Inference **142**, 139–147 (2012)

Grafström, A., Tillé, Y.: Doubly balanced spatial sampling with spreading and restitution of auxiliary totals. Environmetrics **24**, 120–131 (2013)

Grafström, A., Lundström, N.L.P., Schelin, L.: Spatially balanced sampling through the Pivotal method. Biometrics **68**, 514–520 (2012)

Horgan, J.M.: Stratification of skewed populations: a review. Int. Stat. Rev. **74**, 67–76 (2006)

Hung, H.-M., Fuller, W.A.: Regression estimation of crop acreages with transformed landsat data as auxiliary variables. J. Bus. Econ. Stat. **5**, 475–482 (1987)

Jacques, P., Gallego, F.J.: The LUCAS project. The new methodology in the 2005/2006 surveys. Agri-environment Workshop, Belgirate. (http://forum.europa.eu.int/hirc/dsis/landstat/info/data/index.htm) (2005)

Khan, M.G.M., Nand, N., Ahma, N.: Determining the optimum strata boundary points using dynamic programming. Surv. Methodol. **34**, 205–214 (2008)

Kozak, M.: Optimal stratification using random search method in agricultural surveys. Stat. Transit. **5**, 797–806 (2004)

Lohr, S.L.: Multiple-frame surveys. In: Pfeffermann, D., Rao, C.R. (eds.) Handbook of Statistics, Sample Surveys: Design, Methods and Applications, vol. 29A, pp. 71–88. Elsevier, The Netherlands (2009)

Mecatti, F.: A single frame multiplicity estimator for multiple frame surveys. Surv. Methodol. **33**, 151–157 (2007)

Montanari, G.E., Ranalli, G.: Nonparametric model calibration estimation in survey sampling. J. Am. Stat. Assoc. **100**, 1429–1442 (2005)

Nusser, S.M., House, C.C.: Sampling, data collection, and estimation in agricultural Surveys. In: Pfeffermann, D., Rao, C.R. (eds.) Sample Surveys Design, Methods and Applications. Handbook of Statistics, vol. 29A, pp. 471–486. Elsevier, Amsterdam (2009)

Pfeffermann, D.: New important developments in small area estimation. Stat. Sci. **28**, 40–68 (2013)

Postiglione, P., Benedetti, R., Piersimoni, F.: Spatial prediction of agricultural crop yield. In: Benedetti, R., Bee, M., Espa, G., Piersimoni, F. (eds.) Agricultural Survey Methods, pp. 369–387. Wiley, Chichester (2010)

Pratesi, M., Salvati, N.: Small area estimation: the EBLUP estimator based on spatially correlated random area effects. Stat. Methods Appl. **17**, 113–141 (2008)

Rao, J.N.K.: Small Area Estimation. Wiley, Hoboken (2003)

Rao, J.N.K., Yu, M.: Small area estimation by combining time series and cross- sectional data. Can. J. Stat. **22**, 511–528 (1994)

Rogerson, P.A., Delmelle, E.: Optimal sampling design for variables with varying spatial importance. Geogr. Anal. **36**, 177–194 (2004)

Rosén, B.: On sampling with probability proportional to size. J. Stat. Plan. Inference **62**, 159–191 (1997)

Singh, A.: Contributions to Sampling Statistics. Springer, Heidelberg (2014)

Stevens Jr., D.L., Olsen, A.R.: Spatially balanced sampling of natural resources. J. Am. Stat. Assoc. **99**, 262–278 (2004)

Tillé, Y.: Sampling Algorithms. Springer Series in Statistics. Springer, New York (2006)

Tillé, Y.: Ten years of balanced sampling with the cube method: an appraisal. Surv. Methodol. **2**, 215–226 (2011)

Tillé, Y., Favre, A.-C.: Optimal allocation in balanced sampling. Stat. Probab. Lett. **1**, 31–37 (2005)

Vogel, F.A.: The evolution and development of agricultural statistics at the United States Department of Agriculture. J. Official Stat. **11**, 161–180 (1995)

Wang, J.-F., Stein, A., Gao, B.-B., Ge, Y.: A review of spatial sampling. Spat. Stat. **2**, 1–14 (2012)

Wu, C.: Optimal calibration estimators in survey sampling. Biometrika **90**, 937–951 (2003)

Wu, C., Sitter, R.R.: A model calibration approach to using complete auxiliary information from survey data. J. Am. Stat. Assoc. **96**, 185–193 (2001)

A Multi-proportion Randomized Response Model Using the Inverse Sampling

Marcella Polisicchio and Francesco Porro

Abstract This paper describes a new procedure to unbiasedly estimate the proportions of t population groups, which at least one is very small and then it can be considered a rare group. This procedure guarantees the privacy protection of the interviewees, as it is based on an extension of the Warner randomized response model. As the estimation regards rare groups, the sampling design considered is the inverse sampling. Some characteristics of the proposed estimators are investigated.

Keywords Inverse sampling • Randomized response • Sensitive questions

1 Introduction

Sample surveys are fundamental to better understand the society we live in, and the reliability on the results is a crucial point of the whole process. The surveys regarding sensitive issues, stigmatizing attributes, or in general behaviors which are not accepted by the majority of the population, can sometimes embarrass the interviewees, and therefore they can be difficult to perform. When dealing with such personal information, the refusal to respond or even worse, intentional incorrect answers must be taken into account by the researcher, since it is evident that the non-sampling error induced can bear unreliable estimates (for instance, see Cochran 1963).

Warner (1965) introduced an ingenious method that overcomes this issue. He proposed a randomized response technique, which provides an unbiased estimator for the unknown proportion π_A of the persons bearing a stigmatizing characteristic A, with no privacy violation of the interviewee. There is no privacy violation in

M. Polisicchio • F. Porro (✉)
Università degli Studi di Milano-Bicocca, Milano, Italy
e-mail: marcella.polisicchio@unimib.it; francesco.porro1@unimib.it

F. Mecatti et al. (eds.), *Contributions to Sampling Statistics*, Contributions to Statistics, 199
DOI 10.1007/978-3-319-05320-2_13,
© Springer International Publishing Switzerland 2014

the sense that the interviewer receives an answer ("yes" or "no") to the question "Is the statement on the card you drew, true for you?", but he does not know what the statement says. Even if one can believe that nowadays such technique has to be considered overcome, for example, by the computer-assisted self-interview, in the literature there are some studies which prove the contrary (see, for example, Van der Heijden et al. 2000; Van der Heijden and Bockenholt 2008). From the innovative paper of Warner (1965) a lot of improvements have been implemented and a lot of more refined techniques have been developed. Multiple randomized response devices (see Mangat and Singh 1990; Singh 2002; Gjestvang and Singh 2006), modified randomized response techniques (see Mangat 1994; Kuk 1990), models taking into account the probability of lying (see Mangat and Singh 1995), models with one or more scrambled variables (see Gupta et al. 2002; Gjestvang and Singh 2006) and models with unrelated questions (see Singh and Mathur 2006; Pal and Singh 2012)—just to report some examples—are all models or techniques originated by the intuition of Warner. In particular, an interesting extension to the case where the aim is the estimation of the proportions of t-related mutually exclusive population groups, with at least one and at most $t - 1$ of which exhibit sensitive characteristics, is due to Abul-Ela et al. (1967).

The procedure suggested by Warner and the extension of Abul-Ela et al. are based on the "sample proportion" estimator, within a framework where the sample size is fixed a priori. The use of such estimator can have some drawbacks. The first is that, drawing an "unlucky" sample, it can happen that the sample proportion estimator assumes a value equal to zero: in this case the proportion estimate of the group bearing the sensitive attribute depends only on the parameter settings of randomized response device and does not depend on the empirical results. Another situation where the sample proportion estimator can have some troubles is when the sampling costs are expensive, since the drawing of a sample with no elements with the sensitive attribute implies a discard of the sample and a new drawing: this can cause a relevant increasing of the total cost of the survey. The last point is that if the investigated proportion is small (meaning that the survey deals with rare populations) the literature provides more performant estimators: one of these is based on the inverse sampling technique, introduced by Haldane (see Haldane 1945a,b). In such technique, the sample size is not fixed a priori, since the sampling continues until a certain quantity, say k, of elements with the considered attribute are drawn from the population. Many papers in the literature developed the inverse sampling (for further details, see Finney 1949; Best 1974; Mikulski and Smith 1976; Sathe 1977; Sahai 1980; Prasad and Sahai 1982; Pathak and Sathe 1984; Mangat and Singh 1991; Singh and Mathur 2002b; Chaudhuri 2011; Chaudhuri et al. 2011a, just to mention some references).

The aim of this paper is to use the randomized response technique and the inverse sampling together. More in detail, this paper proposes a method to estimate the t proportions of related mutually exclusive population groups, with at least one rare group. This method is based on the inverse sampling technique. Indeed in the literature the randomized response technique and the inverse sampling have been already used together, but as far as the authors know, only to approach the case where one proportion has to be estimated (refer to Singh and Mathur 2002a for further details).

In order to simplify the explanation, the present paper will describe the case where $t = 3$, that is the trinomial case, since the generalization to the case $t > 3$ straightforwardly follows.

The plan of the paper is the following: the next section provides a brief explanation of the randomized response technique of Warner, and the extension of Abul-Ela et al. to the trinomial case. In Sect. 3 the new procedure based on the inverse sampling, for obtaining the proportion estimators is described, and three propositions about their features are stated and proved in detail. Section 4 provides an efficiency comparison with the estimators proposed by Abul-Ela et al. Section 5 is devoted to the estimation of the variances and the covariance of the proposed estimators. In order to improve the proportion estimates, three shrinkage estimators are proposed in Sect. 6. The last section describes some final remarks.

2 The Randomized Response Model of Warner and the Extension to the Trinomial Case of Abul-Ela et al.

Let $\mathbb{S}_1$ be a population, consisting of two mutually exclusive groups. The persons in the first group, say A, bear a sensitive attribute. The aim of the survey is the estimation of the proportion π_A of the group A. Since the direct question "Are you in group A?" can be embarrassing for the interviewee, Warner (1965) proposed a procedure equivalent to the following one. A deck of cards is given to each interviewee. On each card is reported only one of these two statements:

(1) Statement I: "I belong to the group A".
(2) Statement II: "I do not belong to the group A".

The proportion p (with $p \neq 1/2$) of cards with the statement I is known and fixed before the beginning of the survey by the researcher. Obviously, the proportion of cards with the statement II is then $1 - p$. The interviewee is requested to shuffle the deck, to draw a card, and to answer "yes" or "no" if the statement on the card is true or no for him/her, respectively. Since the interviewer does not see the card, he does not know the statement on it, and therefore the privacy of the respondent is not violated. Using this setting, the proportion π_A can be unbiasedly estimated by

$$\hat{\pi}_A = \frac{\hat{\lambda} - (1 - p)}{2p - 1},$$

where $\hat{\lambda}$ is the sample proportion estimator of "yes" answers.

As aforementioned, Abul-Ela et al. extended the Warner procedure, in order to estimate the proportions of t groups: in Abul-Ela et al. (1967) they described in detail the case $t = 3$. Here, a summary of such procedure is reported.

Let $\mathbb{S}_2$ be a population with three exhaustive and mutually exclusive groups (say A, B, and C). Let π_i $(i = 1, 2, 3)$ be the true unknown proportions of the three groups, that is:

$\pi_1 = $ the true proportion of group A in the population;

$\pi_2 = $ the true proportion of group B in the population;

$\pi_3 = $ the true proportion of group C in the population;

with

$$\pi_i \in [0, 1], \quad \text{and} \quad \sum_{i=1}^{3} \pi_i = 1. \tag{1}$$

Since the independent parameters to be estimated are two, from the population two independent non-overlapping random samples with replacement are drawn: the first one with size m_1, the second one with size m_2.

Remark 1. It is worth underlining that here the two sample sizes m_1 and m_2 are fixed a priori by the researcher before the beginning of the survey.

The randomized devices are two decks of cards, one for each sample. There are three kinds of cards in each deck. For the deck i, $(i = 1, 2)$, let

p_{i1} be the proportion of cards with the statement: "I belong to group A";

p_{i2} be the proportion of cards with the statement: "I belong to group B";

p_{i3} be the proportion of cards with the statement: "I belong to group C".

Obviously it holds that

$$p_{ij} \in [0, 1], \quad \text{and} \quad \sum_{j=1}^{3} p_{ij} = 1, \ i = 1, 2. \tag{2}$$

Merely for a technical reason that will be evident in the following, the proportions p_{ij} of the cards in the two decks must satisfy the restriction:

$$(p_{11} - p_{13})(p_{22} - p_{23}) \neq (p_{12} - p_{13})(p_{21} - p_{23}). \tag{3}$$

Abul-Ela et al. (1967) proposed to consider the random variables X_{ir} $(i = 1, 2)$, defined by

$$X_{ir} = \begin{cases} 0 \text{ if the } r\text{-th interviewee in sample } i \text{ says "no"} \\ 1 \text{ if the } r\text{-th interviewee in sample } i \text{ says "yes"} \end{cases}$$

where $r \in \{1, 2, \ldots, m_1\}$ if $i = 1$, and $r \in \{1, 2, \ldots, m_2\}$ if $i = 2$. The probabilities of "yes" answer for the r-th interviewee in the two samples are

$$\mathbb{P}(X_{1r} = 1) = \mathbb{P}(X_{1r} = 1|A)\mathbb{P}(A) + \mathbb{P}(X_{1r} = 1|B)\mathbb{P}(B) + \mathbb{P}(X_{1r} = 1|C)\mathbb{P}(C)$$
$$= p_{11}\pi_1 + p_{12}\pi_2 + p_{13}\pi_3,$$

and

$$\mathbb{P}(X_{2r} = 1) = \mathbb{P}(X_{2r} = 1|A)\mathbb{P}(A) + \mathbb{P}(X_{2r} = 1|B)\mathbb{P}(B) + \mathbb{P}(X_{2r} = 1|C)\mathbb{P}(C)$$
$$= p_{21}\pi_1 + p_{22}\pi_2 + p_{23}\pi_3,$$

where $\mathbb{P}(X_{ir} = 1|A)$ denotes the probability that the r-th interviewee in the sample i $(i = 1, 2)$ says "yes", given that he/she belongs to group A and $\mathbb{P}(A)$ is the probability that a person selected belongs to the group A. For the restrictions (2) and (1), the above probabilities become

$$\mathbb{P}(X_{1r} = 1) = (p_{11} - p_{13})\pi_1 + (p_{12} - p_{13})\pi_2 + p_{13},$$

and

$$\mathbb{P}(X_{2r} = 1) = (p_{21} - p_{23})\pi_1 + (p_{22} - p_{23})\pi_2 + p_{23}.$$

Denoting $\mathbb{P}(X_{1r} = 1)$ by λ_1 and $\mathbb{P}(X_{2r} = 1)$ by λ_2, from the equations

$$\begin{cases} \lambda_1 = (p_{11} - p_{13})\pi_1 + (p_{12} - p_{13})\pi_2 + p_{13} \\ \lambda_2 = (p_{21} - p_{23})\pi_1 + (p_{22} - p_{23})\pi_2 + p_{23} \\ \pi_3 = 1 - (\pi_1 + \pi_2) \end{cases} \tag{4}$$

it derives that

$$\begin{cases} \pi_1 = C \cdot [(\lambda_1 - p_{13})(p_{22} - p_{23}) - (\lambda_2 - p_{23})(p_{12} - p_{13})] \\ \pi_2 = -C \cdot [(\lambda_1 - p_{13})(p_{21} - p_{23}) - (\lambda_2 - p_{23})(p_{11} - p_{13})] \\ \pi_3 = 1 - C\,[(\lambda_1 - p_{13})(p_{22} - p_{21}) - (\lambda_2 - p_{23})(p_{12} - p_{11})], \end{cases} \tag{5}$$

where the constant C is given by

$$C = [(p_{11} - p_{13})(p_{22} - p_{23}) - (p_{12} - p_{13})(p_{21} - p_{23})]^{-1}. \tag{6}$$

Remark 2. The restriction (3) guarantees that the constant C is well defined.

The equations in (5) describe the unknown proportions π_1, π_2 and π_3 in terms of the quantities λ_1 and λ_2, which can be estimated using the sample proportion estimators. Then, denoting by Y_1 the number of "yes" answers in the first sample, and by Y_2 the number of "yes" answers in the second sample, the random variables

$$\hat{\lambda}_1^A = \frac{Y_1}{m_1} \qquad \text{and} \qquad \hat{\lambda}_2^A = \frac{Y_2}{m_2} \tag{7}$$

can be used to estimate the probabilities λ_1 and λ_2. Since the sampling is with replacement, the random variables Y_1 and Y_2 have binomial distribution, with parameters (λ_1, m_1) and (λ_2, m_2), respectively. Replacing these estimators in (5), Abul-Ela et al. obtained the following estimators for the proportions π_i ($i = 1, 2, 3$):

$$
\begin{cases}
\hat{\pi}_1^A = C \cdot \left[\left(\hat{\lambda}_1^A - p_{13} \right) (p_{22} - p_{23}) - \left(\hat{\lambda}_2^A - p_{23} \right) (p_{12} - p_{13}) \right] \\
\hat{\pi}_2^A = -C \cdot \left[\left(\hat{\lambda}_1^A - p_{13} \right) (p_{21} - p_{23}) - \left(\hat{\lambda}_2^A - p_{23} \right) (p_{11} - p_{13}) \right] \\
\hat{\pi}_3^A = 1 - C \left[\left(\hat{\lambda}_1^A - p_{13} \right) (p_{22} - p_{21}) - \left(\hat{\lambda}_2^A - p_{23} \right) (p_{12} - p_{11}) \right],
\end{cases}
\tag{8}
$$

where the last equation is obtained from $\hat{\pi}_3^A = 1 - (\hat{\pi}_1^A + \hat{\pi}_2^A)$. Such estimators are unbiased and they have some interesting features (see Abul-Ela et al. 1967; Chaudhuri 2011 for further details). It is worth noting that the variances of the estimators in (8) can be directly evaluated by

$$
\begin{aligned}
\text{var}(\hat{\pi}_1^A) &= C^2 \left[(p_{22} - p_{23})^2 \, \varphi_1^2 + (p_{12} - p_{13})^2 \, \varphi_2^2 \right] \\
\text{var}(\hat{\pi}_2^A) &= C^2 \left[(p_{21} - p_{23})^2 \, \varphi_1^2 + (p_{11} - p_{13})^2 \, \varphi_2^2 \right] \\
\text{var}(\hat{\pi}_3^A) &= C^2 \left[(p_{22} - p_{21})^2 \varphi_1^2 + (p_{12} - p_{11})^2 \varphi_2^2 \right],
\end{aligned}
\tag{9}
$$

where φ_i^2 denotes the variance of the sample proportion estimators $\hat{\lambda}_i^A$ defined in (7):

$$\varphi_i^2 = \text{var}(\hat{\lambda}_i^A) = \frac{\lambda_i (1 - \lambda_i)}{m_i}, \quad i = 1, 2. \tag{10}$$

3   The Introduction of the Inverse Sampling Technique

This section describes a procedure to estimate the proportions π_i ($i = 1, 2, 3$). As far as the authors know, there is no mention in the literature of such procedure, then it can be considered innovative.

As stated in the introduction, such procedure is based on the inverse sampling technique, where the sample size is not fixed a priori. The needed setting is the same one illustrated in the previous section: the population, the two decks with proportions of the cards, and the randomized response technique are exactly the same ones previously described.

The crucial difference from the procedure of Abul-Ela et al. is that, here, the sizes of the two samples are not fixed a priori. The drawing (with replacement) of persons in the first sample continues until that k_1 "yes" answers are obtained, and the drawing (with replacement) of persons in the second sample finishes when k_2 "yes" answer are reported. The values of the parameters k_1 and k_2 are decided by the researcher before the beginning of the survey: the most interesting situations are when k_1 and k_2 are greater or equal to 2. In such cases, the sample sizes, say N_1 and N_2, are random variables with negative binomial distribution, with parameters (λ_1, k_1) and (λ_2, k_2), respectively. In the proposed procedure, only the sampling with replacement is examined. The case where only distinct units are considered poses further difficulties, and hence it will not be treated in this paper (for further details, see Chaudhuri 2011; Chaudhuri et al. 2011b).

The criterion for fixing the value of the parameter k_i ($i = 1, 2$) depends on many aspects, for example, the population size, the sampling costs, the time and resources needed for the sampling procedure. For this reason, the determination of these parameters is an important issue and it varies from case to case. A possible criterion, using data from previous surveys (if available) is to select the value of the parameter k_i ($i = 1, 2$) as the product between a reasonable (and cost-bearable) expected sample size and the previous estimate of λ_i ($i = 1, 2$). In such way the sample size is random, but at least the expected sample size can be kept under control, since the expectation of the size for the sample i is given by k_i / λ_i ($i = 1, 2$).

Using the notation of the previous section, in order to obtain estimators for the proportions π_i ($i = 1, 2, 3$), two estimators of λ_1 and λ_2 are needed.

In the literature, when dealing with the inverse sampling scheme, the most used estimator for the proportion of a group in the population is given by the ratio between the prefixed number of "successes" that concludes the sampling minus one and the sample size minus one. This estimator is unbiased and its variance can be evaluated as the sum of a series. These and further important and useful features are discussed and proved in Haldane (1945a,b), Best (1974), and Chaudhuri (2011).

In the present framework dealing with the inverse sampling, the sample proportion estimators used by Abul-Ela et al. cannot be utilized. Instead of them, the following two estimators can be used:

$$\hat{\lambda}_1 = \frac{k_1 - 1}{N_1 - 1} \qquad \text{and} \qquad \hat{\lambda}_2 = \frac{k_2 - 1}{N_2 - 1}, \tag{11}$$

where k_1 and k_2 are the fixed numbers of "yes" answers which conclude the samplings and N_1 and N_2 are the random variables representing the sizes of the two samples. Following the same approach of Abul-Ela et al., the plug-in estimators of the unknown proportions π_i ($i = 1, 2, 3$) can then be obtained as:

$$\begin{cases} \hat{\pi}_1 = C \cdot \left[\left(\hat{\lambda}_1 - p_{13} \right) (p_{22} - p_{23}) - \left(\hat{\lambda}_2 - p_{23} \right) (p_{12} - p_{13}) \right] \\ \hat{\pi}_2 = -C \cdot \left[\left(\hat{\lambda}_1 - p_{13} \right) (p_{21} - p_{23}) - \left(\hat{\lambda}_2 - p_{23} \right) (p_{11} - p_{13}) \right] \\ \hat{\pi}_3 = 1 - C \left[\left(\hat{\lambda}_1 - p_{13} \right) (p_{22} - p_{21}) - \left(\hat{\lambda}_2 - p_{23} \right) (p_{12} - p_{11}) \right], \end{cases} \tag{12}$$

where the estimators $\hat{\lambda}_1$, $\hat{\lambda}_2$ are defined in (11), and the third estimator $\hat{\pi}_3$ is derived as $\hat{\pi}_3 = 1 - (\hat{\pi}_1 + \hat{\pi}_2)$.

These estimators have several interesting characteristics: some of them are stated in the following propositions.

Proposition 1. *The estimators defined in (12) are unbiased for the unknown proportions π_i $(i = 1, 2, 3)$.*

Proof. For the first estimator $\hat{\pi}_1$ it holds that

$$\mathbb{E}(\hat{\pi}_1) = \mathbb{E}\left[C \left[\left(\hat{\lambda}_1 - p_{13} \right) (p_{22} - p_{23}) - \left(\hat{\lambda}_2 - p_{23} \right) (p_{12} - p_{13}) \right] \right]$$

$$= C \left[(p_{22} - p_{23}) \left[\mathbb{E}\left(\hat{\lambda}_1 \right) - p_{13} \right] - (p_{12} - p_{13}) \left[\mathbb{E}\left(\hat{\lambda}_2 \right) - p_{23} \right] \right].$$

Since $\hat{\lambda}_1$ and $\hat{\lambda}_2$ are unbiased (as proved in Haldane 1945b), it follows:

$$\mathbb{E}(\hat{\pi}_1) = C \left[(p_{22} - p_{23}) (\lambda_1 - p_{13}) - (p_{12} - p_{13}) (\lambda_2 - p_{23}) \right],$$

hence, using (4) and (6):

$$\mathbb{E}(\hat{\pi}_1) = C \left[(p_{22} - p_{23})[(p_{11} - p_{13})\pi_1 + (p_{12} - p_{13})\pi_2] + \right.$$

$$\left. -(p_{12} - p_{13})[(p_{21} - p_{23})\pi_1 + (p_{22} - p_{23})\pi_2] \right]$$

$$= \pi_1 C \left[(p_{22} - p_{23})(p_{11} - p_{13}) - (p_{12} - p_{13})(p_{21} - p_{23}) \right]$$

$$= \pi_1.$$

Through an analogous procedure it can be shown that also $\hat{\pi}_2$ is unbiased:

$$\mathbb{E}(\hat{\pi}_2) = -C \left[(p_{21} - p_{23}) \left[\mathbb{E}\left(\hat{\lambda}_1 \right) - p_{13} \right] - (p_{11} - p_{13}) \left[\mathbb{E}\left(\hat{\lambda}_2 \right) - p_{23} \right] \right] = \pi_2,$$

while for the last estimator $\hat{\pi}_3$ it holds:

$$\mathbb{E}(\hat{\pi}_3) = \mathbb{E}[1 - (\hat{\pi}_1 + \hat{\pi}_1)] = 1 - \pi_1 - \pi_2 = \pi_3.$$

Remark 3. By construction, the two estimators $\hat{\lambda}_1$ and $\hat{\lambda}_2$ are independent, as the first one is referred to the first sample, while the latter is referred to the second sample and the two samples are independent and non-overlapping. On the contrary, the two estimators $\hat{\pi}_1$ and $\hat{\pi}_2$ are not independent since each of them depends on both $\hat{\lambda}_1$ and $\hat{\lambda}_2$.

Now, the variance of the estimators defined in (12) can be evaluated, as the following proposition states.

Proposition 2. *The variances of the estimators defined in (12) are:*

$$\text{var}(\hat{\pi}_1) = C^2\left[(p_{22} - p_{23})^2\,\delta_1^2 + (p_{12} - p_{13})^2\,\delta_2^2\right]$$
$$\text{var}(\hat{\pi}_2) = C^2\left[(p_{21} - p_{23})^2\,\delta_1^2 + (p_{11} - p_{13})^2\,\delta_2^2\right] \quad\quad (13)$$
$$\text{var}(\hat{\pi}_3) = C^2\left[(p_{22} - p_{21})^2\delta_1^2 + (p_{12} - p_{11})^2\delta_2^2\right],$$

where δ_i^2 denotes the variance of the estimator $\hat{\lambda}_i$ defined in (11):

$$\delta_i^2 = \text{var}(\hat{\lambda}_i) = \lambda_i^2 \sum_{r=1}^{+\infty} \binom{k_i + r - 1}{r}^{-1} (1 - \lambda_i)^r, \quad i = 1, 2. \quad\quad (14)$$

Proof. For the first estimator $\hat{\pi}_1$, from the definition (12) it follows that

$$\text{var}(\hat{\pi}_1) = C^2 \text{var}\left[\left(\hat{\lambda}_1 - p_{13}\right)(p_{22} - p_{23}) - \left(\hat{\lambda}_2 - p_{23}\right)(p_{12} - p_{13})\right].$$

As stated in the Remark 3, since $\hat{\lambda}_1$ and $\hat{\lambda}_2$ are independent, it holds that

$$\text{var}(\hat{\pi}_1) = C^2\left[(p_{22} - p_{23})^2 \text{var}\left(\hat{\lambda}_1\right) + (p_{12} - p_{13})^2 \text{var}\left(\hat{\lambda}_2\right)\right]. \quad\quad (15)$$

As briefly mentioned, Best (1974) proved that the variances δ_i^2 of the estimators $\hat{\lambda}_i$ defined in (11) are given by the formula (14), therefore replacing of such expressions in (15) concludes the proof for $\hat{\pi}_1$. With a similar procedure the variance of $\hat{\pi}_2$ and $\hat{\pi}_3$ can be evaluated as in (13).

In order to have a graphical overview on the behaviour of these variances when the parameters are modified, in Figs. 1 and 2 some graphs are drawn. In such graphs, the variance of $\hat{\pi}_1$ is considered, but the analogous remarks hold for $\hat{\pi}_2$. In all the figures, the settings of the randomized response devices (that is, the proportions of the cards in the two decks) are equal and given by

$$
\begin{aligned}
p_{11} &= 0.1 & p_{21} &= 0.2 \\
p_{12} &= 0.2 & p_{22} &= 0.5 \\
p_{13} &= 0.7 & p_{23} &= 0.3.
\end{aligned}
$$

Figure 1 shows how the variance of the estimator $\hat{\pi}_1$ changes as the values of the parameters k_1 (on the left) and k_2 (on the right) are modified. In both the graphs, the value of π_2 is fixed and equal to 0.3, while the value of π_1 is reported in the legenda, and the value of π_3 equals $1 - (\pi_1 + \pi_2)$. The difference between the two graphs is that in the one on the left, the value of k_2 is fixed and equal to 5, while in the other one k_1 is fixed and equal to 5.

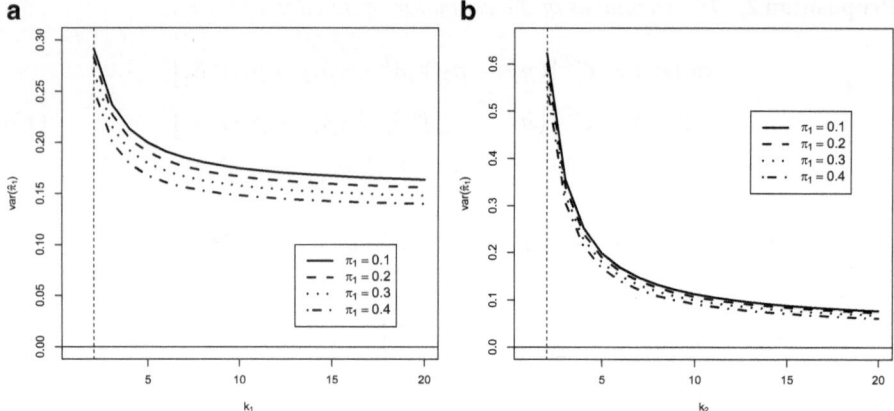

Fig. 1 The behaviour of the variance of $\hat{\pi}_1$, when k_1 (**a**) and when k_2 (**b**) varies, for different value of π_1

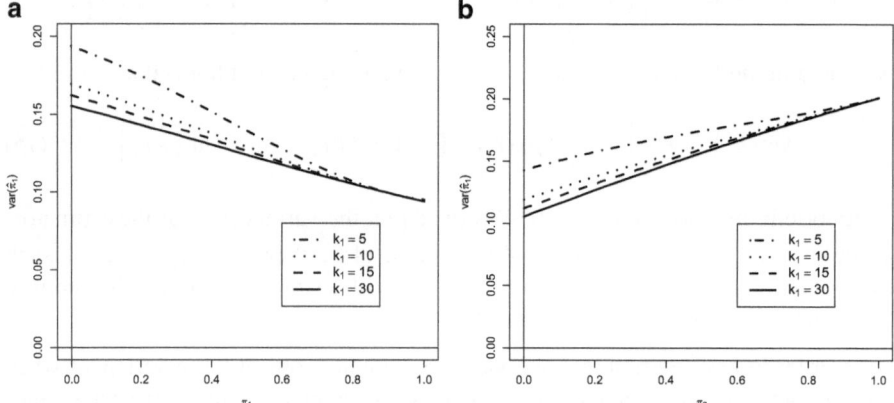

Fig. 2 The behavior of the variance of $\hat{\pi}_1$, when π_1 (**a**) and when π_2 (**b**) varies, for different value of k_1

The graph on the left in Fig. 1 shows that as k_1 increases, the variance of $\hat{\pi}_1$ decreases: such behaviour is not surprising, since a large k_1 means a large size of the first sample and then, at least intuitively, less variability of the estimators. The same arguments can be applied for the graph on the right side: here, moreover it is evident that not only $\hat{\pi}_2$, but also $\hat{\pi}_1$, as aspected, depends on the value of the parameter k_2. The different rate of decrease of $\text{var}(\hat{\pi}_1)$ in the two graphs can be explained by the different settings of the two decks of cards. Figure 2 shows how the variance of $\hat{\pi}_1$ depends on the true values of the proportions π_1 and π_2. In the graph on the left π_2 is set equal to 0.3, while in the other one the value of π_1 is 0.3.

Remark 4. The variances in the Proposition 2 are not easy to manage and for some values of the parameters, they are difficult to be evaluated, even only numerically. To overcome this issue, the finding of an upper bound for the variances may be useful, especially in the applications.

In Singh and Mathur (2005) eleven upper bounds for the variance of the estimators defined in (11) are compared: the main result is that the following upper bound proposed in Sahai (1983)

$$W_1(k) = \frac{\lambda}{6k} \left[\sqrt{A^2 - 12k\lambda B} - A \right]$$

where

λ is the unknown proportion of the "successes";

k is the number of "successes" that concludes the sampling procedure;

$A = k^2 + k(3\lambda - 1) - 3\lambda(1 - \lambda) - \dfrac{6(1 - \lambda)^2}{k + 1}$;

$B = (1 - \lambda) \left[\dfrac{(1 - \lambda)(k - 1)}{k + 1} - (k + 2) \right]$,

is the closest to the exact variance for smaller size for k (i.e. $k \leq 7$). For large k (i.e $k \geq 8$), the following upper bound, due to Pathak and Sathe (1984),

$$W_2(k) = \frac{\lambda^2(1 - \lambda)}{k} \left[1 + \frac{2(1 - \lambda)}{k - 2} \times \right.$$

$$\left. \times \left(1 - \frac{6\lambda}{k - 3(1 - \lambda) + 1 + \sqrt{(k - 5(1 - \lambda) + 1)^2 + 16\lambda(1 - \lambda)}} \right) \right]$$

is the best choice.

Applying such result to the estimators defined in (12), the three upper bounds U_i for their variances can be obtained, since:

$$\mathrm{var}(\hat{\pi}_1) \leq U_1 = C^2 \left[(p_{22} - p_{23})^2 \, W(k_1) + (p_{12} - p_{13})^2 \, W(k_2) \right]$$

$$\mathrm{var}(\hat{\pi}_2) \leq U_2 = C^2 \left[(p_{21} - p_{23})^2 \, W(k_1) + (p_{11} - p_{13})^2 \, W(k_2) \right] \qquad (16)$$

$$\mathrm{var}(\hat{\pi}_3) \leq U_3 = C^2 \left[(p_{22} - p_{21})^2 \, W(k_1) + (p_{12} - p_{11})^2 \, W(k_2) \right],$$

where

$$W(k) = \begin{cases} W_1(k) & \text{for } k \leq 7 \\ W_2(k) & \text{for } k \geq 8. \end{cases}$$

As described in the Remark 3 the two estimators $\hat{\pi}_1$ and $\hat{\pi}_2$ are not independent since each of them depends on both $\hat{\lambda}_1$ and $\hat{\lambda}_2$: for this reason the evaluation of the covariance between $\hat{\pi}_1$ and $\hat{\pi}_2$ needs an investigation.

Proposition 3. *The covariance between the estimators $\hat{\pi}_1$ and $\hat{\pi}_2$ defined in (12) is*

$$\text{cov}(\hat{\pi}_1, \hat{\pi}_2) = C^2[(p_{22} - p_{23})(p_{23} - p_{21})\delta_1^2 + (p_{12} - p_{13})(p_{13} - p_{11})\delta_2^2].$$

Proof. In the previous proposition it has been proved that

$$\text{var}(\hat{\pi}_3) = \text{var}(\hat{\pi}_1 + \hat{\pi}_2) = C^2\left[(p_{22} - p_{21})^2\delta_1^2 + (p_{12} - p_{11})^2\delta_2^2\right]$$

where δ_i^2 denotes the variance of the estimator $\hat{\lambda}_i$ $(i = 1, 2)$, defined in (11). By the definition of the variance of the sum of two random variables, it holds that

$$C^2\left[(p_{22} - p_{21})^2\delta_1^2 + (p_{12} - p_{11})^2\delta_2^2\right] = \text{var}(\hat{\pi}_1) + \text{var}(\hat{\pi}_2) + 2\,\text{cov}(\hat{\pi}_1, \hat{\pi}_2).$$

Recalling from the previous proposition that

$$\text{var}(\hat{\pi}_1) = C^2\left[(p_{22} - p_{23})^2\,\delta_1^2 + (p_{12} - p_{13})^2\,\delta_2^2\right]$$

and

$$\text{var}(\hat{\pi}_2) = C^2\left[(p_{21} - p_{23})^2\,\delta_1^2 + (p_{11} - p_{13})^2\,\delta_2^2\right],$$

after some algebra it follows that

$$\text{cov}(\hat{\pi}_1, \hat{\pi}_2) = C^2[(p_{22} - p_{23})(p_{23} - p_{21})\delta_1^2 + (p_{12} - p_{13})(p_{13} - p_{11})\delta_2^2].$$

Remark 5. As before with the variances, the expression of the covariance can be difficult to manage. However, using the aforementioned upper bounds U_1 and U_2 and the Cauchy–Schwartz inequality, it is possible to determinate the following interval, where the covariance lies in:

$$\left[-\sqrt{U_1 U_2}\ ;\ \sqrt{U_1 U_2}\right].$$

The estimators $\hat{\pi}_i$ $(i = 1, 2, 3)$ proposed in this section have beyond the features stated in the previous propositions at least two other important characteristics. The first one is that the values of these estimators always depend on the sample information: this basic issue is not true for the estimators proposed by Abul-Ela et al., since if a sample contains no "yes" answers, the associated estimators sample proportion λ_i^A assumes value zero, and therefore the estimators $\hat{\pi}_i$ $(i = 1, 2, 3)$ capture no information from this sample. A possible solution to this drawback can be the discard of such "unlucky" sample, but if the sampling costs are not negligible, such operation can largely increase the total cost of the survey, and therefore in particular situations cannot be performed. Using the procedure proposed in this paper, such situation cannot occur, since the estimators $\hat{\lambda}_i$ $(i = 1, 2)$ cannot be equal to zero.

The last remark is about the impression that the proposed estimators $\hat{\pi}_i$ $(i = 1, 2, 3)$ do not really differ so much from the ones introduced by Abul-Ela et al. From the formal point of view such idea can arise, as the expressions (12) look very similar to the ones in (8), but indeed the two procedures deal with two very different experiments: the one of Abul-Ela et al. is based on sample sizes prefixed, while the procedure proposed in this paper has random sample sizes.

4 An Efficiency Comparison

Since the proposed procedure provides three new estimators for the proportions π_i $(i = 1, 2, 3)$, a comparison with the ones proposed by Abul-Ela et al. (1967) should be interesting. All of these estimators are unbiased, therefore they can be compared, through an analysis of their variances. The expressions of $\text{var}(\hat{\pi}_i^A)$ $(i = 1, 2, 3)$ are reported in (9), while the $\text{var}(\hat{\pi}_i)$ $(i = 1, 2, 3)$ are described in (13).

This section compares the estimators $\hat{\pi}_1^A$ and $\hat{\pi}_1$, but the same investigation can be performed for the other estimators, obtaining analogous results.

Since the estimators $\hat{\pi}_1^A$ and $\hat{\pi}_1$ refer to two different sample designs, a reasonable comparison of their variance can be performed, assuming that the expected sample size is the same in the two sample designs.

Figure 3 shows on the left the variance of the estimator $\hat{\pi}_1^A$, and on the right, the variance of the estimator $\hat{\pi}_1$, in function of the true value of the proportion π_1.

In both the figures, the setting of the randomized response devices (i.e. the cards in the two decks) is the same one used for the figures in the previous section, and π_2 is equal to 0.3.

In the figure on the right, each curve corresponds to a particular choice of the parameters. The value of the parameter k_2 is fixed and equal to 6, while k_1 assumes three values: 3 (solid line), 10 (dashed line) and 100 (dotted line). As noted in the previous section, the variance curves are ordered: the curve corresponding to a large k_1 lies below the curve corresponding to a small k_1.

The variance of the estimator $\hat{\pi}_1^A$ of Abul-Ela et al. does not depend on k_1 and k_2, but it depends on the two sample sizes m_1 and m_2. In order to design three curves comparable to those of $\hat{\pi}_1$, the values of m_1 and m_2 have been set, applying the following rule:

$$m_1 = \frac{k_1}{\lambda_1} \qquad \text{and} \qquad m_2 = \frac{k_2}{\lambda_2}. \qquad (17)$$

In other words, the values of m_1 in the first curve (in solid) on the left side are calculated as the ratio between $k_1 = 3$ and the value of λ_1, obtained using formula (4), which varies in function of π_1. In such way, as mentioned in the previous

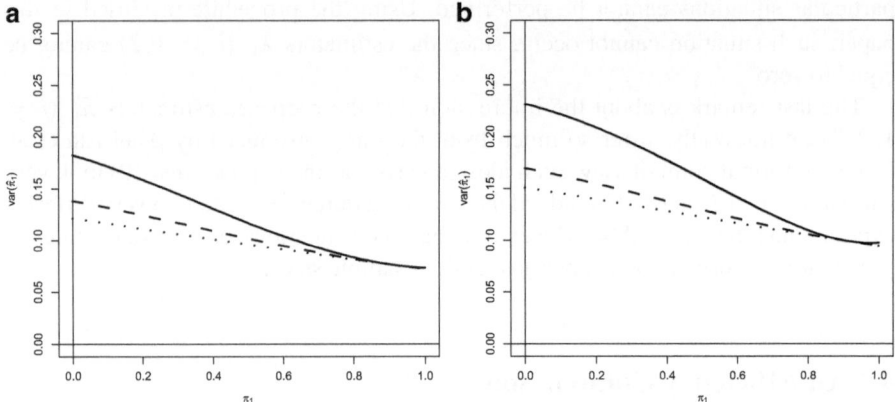

Fig. 3 The behaviour of the variance of $\hat{\pi}_1^A$ (**a**) and of $\hat{\pi}_1$ (**b**), for different values of the parameters, when π_1 varies

sections, for a fixed π_1, the parameters of two corresponding curves in the two figures guarantee that the expectations of the sample size in the two experiments are equal.

The comparison of the curves makes clear basically two points: first, the variance of $\hat{\pi}_1^A$ is always smaller than the variance of $\hat{\pi}_1$; second, the behaviour of the variance curves of $\hat{\pi}_1$ and $\hat{\pi}_1^A$, as the true proportion π_1 varies, is quite similar.

The first point is not surprising: the estimators proposed in this paper are based on the inverse sampling, which outperforms other sampling designs in particular situations, but it does not reduce the uncertainty and therefore the variability of the estimators.

Also the second point can be explained. The relationship between the variance of $\hat{\lambda}_i$ (denoted by δ_i^2) and the variance of $\hat{\lambda}_i^A$ (denoted by φ_i^2) for ($i = 1, 2$) can be made explicit by

$$
\begin{aligned}
\delta_i^2 &= \text{var}(\hat{\lambda}_i) \\
&= \lambda_i^2 \sum_{r=1}^{+\infty} \binom{k_i + r - 1}{r}^{-1} (1 - \lambda_i)^r \\
&= \frac{\lambda_i^2 (1 - \lambda_i)}{k_i} + \lambda_i^2 \sum_{r=2}^{+\infty} \binom{k_i + r - 1}{r}^{-1} (1 - \lambda_i)^r \\
&= \text{var}(\hat{\lambda}_i^A) \frac{\lambda_i m_i}{k_i} + \lambda_i^2 \sum_{r=2}^{+\infty} \binom{k_i + r - 1}{r}^{-1} (1 - \lambda_i)^r \\
&= \varphi_i^2 \frac{\lambda_i m_i}{k_i} + \xi(k_i, \lambda_i).
\end{aligned}
\tag{18}
$$

Using the formula (18), the variance of $\hat{\pi}_1$ becomes:

$$\mathrm{var}(\hat{\pi}_1) = C^2[(p_{22} - p_{23})^2\delta_1^2 + (p_{12} - p_{13})^2\delta_2^2]$$

$$= C^2\left[(p_{22} - p_{23})^2\left(\varphi_1^2\frac{\lambda_1 m_1}{k_1} + \xi(k_1, \lambda_1)\right) + \right.$$

$$\left. + (p_{12} - p_{13})^2\left(\varphi_2^2\frac{\lambda_2 m_2}{k_2} + \xi(k_2, \lambda_2)\right)\right]$$

$$= C^2\left[(p_{22} - p_{23})^2\left(\varphi_1^2\frac{\lambda_1 m_1}{k_1}\right) + (p_{12} - p_{13})^2\left(\varphi_2^2\frac{\lambda_2 m_2}{k_2}\right)\right] +$$

$$+ C^2\left[(p_{22} - p_{23})^2\xi(k_1, \lambda_1) + (p_{12} - p_{13})^2\xi(k_2, \lambda_2)\right].$$

If the parameters m_1 and m_2 are chosen using the rule (17), then the $\mathrm{var}(\hat{\pi}_1)$ assumes the form

$$\mathrm{var}(\hat{\pi}_1) = \mathrm{var}(\hat{\pi}_1^A) + C^2\left[(p_{22} - p_{23})^2\xi(k_1, \lambda_1) + (p_{12} - p_{13})^2\xi(k_2, \lambda_2)\right].$$

It is worth noting that the quantity ξ introduced in (18) is the sum

$$\xi(k_i, \lambda_i) = \lambda_i^2\sum_{r=2}^{+\infty}\binom{k_i + r - 1}{r}^{-1}(1 - \lambda_i)^r$$

$$= \frac{2!\lambda_i^2(1 - \lambda_i)^2}{k_i(k_i + 1)} + \frac{3!\lambda_i^2(1 - \lambda_i)^3}{k_i(k_i + 1)(k_i + 2)}$$

$$+ \frac{4!\lambda_i^2(1 - \lambda_i)^4}{k_i(k_i + 1)(k_i + 2)(k_i + 3)} + \cdots$$

which decreases as k_i increases. This behaviour of ξ is the reason why in Fig. 3 the difference between $\mathrm{var}(\hat{\pi}_1)$ and $\mathrm{var}(\hat{\pi}_1^A)$ reduces, as k_1 increases.

The comparison highlights that the proposed estimators can be a valid alternative to those introduced by Abul-Ela et al., in sample surveys dealing with rare groups.

5 Unbiased Estimators for the Variances and the Covariance

In some cases, it is useful to have at disposal an estimator for the variances of $\hat{\pi}_i$ ($i = 1, 2, 3$). The following formula provides an unbiased estimator of the variance δ_i^2 ($i = 1, 2$) of the estimators defined in (11) (see Sukhatme et al. 1984 for further details):

$$\widehat{\mathrm{var}(\hat{\lambda}_i)} = \widehat{\delta_i^2} = \frac{\hat{\lambda}_i(1 - \hat{\lambda}_i)}{N_i - 2}, \qquad \text{with } N_i > 2, \qquad i = 1, 2.$$

Using this result, the following estimators are therefore unbiased estimators for the variances $\mathrm{var}(\hat{\pi}_i)$ ($i = 1, 2, 3$):

$$\widehat{\mathrm{var}(\hat{\pi}_1)} = C^2 \left[(p_{22} - p_{23})^2 \, \widehat{\delta_1^2} + (p_{12} - p_{13})^2 \, \widehat{\delta_2^2} \right]$$

$$\widehat{\mathrm{var}(\hat{\pi}_2)} = C^2 \left[(p_{21} - p_{23})^2 \, \widehat{\delta_1^2} + (p_{11} - p_{13})^2 \, \widehat{\delta_2^2} \right] \qquad (19)$$

$$\widehat{\mathrm{var}(\hat{\pi}_3)} = C^2 \left[(p_{22} - p_{21})^2 \widehat{\delta_1^2} + (p_{12} - p_{11})^2 \widehat{\delta_2^2} \right].$$

Analogously, an unbiased estimator for the $\mathrm{cov}(\hat{\pi}_1, \hat{\pi}_2)$ is instead given by:

$$\widehat{\mathrm{cov}(\hat{\pi}_1, \hat{\pi}_2)} = C^2 [(p_{22} - p_{23})(p_{23} - p_{21})\widehat{\delta_1^2} + (p_{12} - p_{13})(p_{13} - p_{11})\widehat{\delta_2^2}]. \qquad (20)$$

A detailed analysis of the features of the estimators defined in (19) and (20) is an interesting and promising subject, but it is out of the scope of the present paper.

6 A Shrinkage Estimator

This section provides three shrinkage estimators for the proportions π_1, π_2 and π_3 described in the previous sections. The basic idea is to reduce the support of the estimators defined in (12), discarding the inadmissible values. Since the considered estimators are estimators of proportions, the values outside of the interval $[0, 1]$ have to be removed. A drawback of such kind of shrinkage is that it transforms unbiased estimators in estimators that can be biased: the balance between such defect and the advantage deriving by the modification of the range has to be evaluated by the researcher, taking into account the final scope of the survey.

The shrinkage method presented in this section provides three new estimators $\hat{\pi}_1^S$, $\hat{\pi}_2^S$ and $\hat{\pi}_3^S$ such that:

- The range of each $\hat{\pi}_i^S$ is the interval $[0, 1]$.
- They sum to 1: $\sum_{i=1}^{3} \hat{\pi}_i^S = 1$.

To simplify the notation, according to the definitions in (12) the following five sets are defined:

$$L_1 = \{(\hat{\lambda}_1, \hat{\lambda}_2) \in]0, 1]^2 : \hat{\pi}_1(\hat{\lambda}_1, \hat{\lambda}_2) \leq 0\};$$

$$L_2 = \{(\hat{\lambda}_1, \hat{\lambda}_2) \in]0, 1]^2 : \hat{\pi}_2(\hat{\lambda}_1, \hat{\lambda}_2) \leq 0\};$$

$$Z_1 = \{(\hat{\lambda}_1, \hat{\lambda}_2) \in]0, 1]^2 : \hat{\pi}_1(\hat{\lambda}_1, \hat{\lambda}_2) \geq 1\};$$

$$Z_2 = \{(\hat{\lambda}_1, \hat{\lambda}_2) \in]0, 1]^2 : \hat{\pi}_2(\hat{\lambda}_1, \hat{\lambda}_2) \geq 1\};$$

$$R = \{(\hat{\lambda}_1, \hat{\lambda}_2) \in]0, 1]^2 : \hat{\pi}_1(\hat{\lambda}_1, \hat{\lambda}_2) + \hat{\pi}_2(\hat{\lambda}_1, \hat{\lambda}_2) > 1\}.$$

As in the previous sections, the value of the last estimator $\hat{\pi}_3$ is obtained at the end, using the values assumed by the others.

The shrinkage procedure is based on the assumption that the first group is the most relevant or significant, therefore the corresponding proportion estimator $\hat{\pi}_1$ has a more important role than the others. For this reason, as importance as possible is given to the empirical evidence regarding the first unknown proportion π_1 of group A. This is realized by the shrinkage, which does not modify the value of $\hat{\pi}_1$ whether it is admissible, hence:

$$\hat{\pi}_1^S(\hat{\lambda}_1, \hat{\lambda}_2) = \begin{cases} 0 & \text{for } (\hat{\lambda}_1, \hat{\lambda}_2) \in L_1 \\ 1 & \text{for } (\hat{\lambda}_1, \hat{\lambda}_2) \in Z_1 \\ \hat{\pi}_1(\hat{\lambda}_1, \hat{\lambda}_2) & \text{otherwise.} \end{cases}$$

The value of $\hat{\pi}_2$ is not modified in only one case, otherwise it changes as indicated below:

$$\hat{\pi}_2^S(\hat{\lambda}_1, \hat{\lambda}_2) = \begin{cases} 1 & \text{for } (\hat{\lambda}_1, \hat{\lambda}_2) \in L_1 \cap Z_2 \\ \hat{\pi}_2(\hat{\lambda}_1, \hat{\lambda}_2) & \text{for } (\hat{\lambda}_1, \hat{\lambda}_2) \in (R \cup L_2 \cup Z_2)^C \\ 1 - \hat{\pi}_1^S(\hat{\lambda}_1, \hat{\lambda}_2) & \text{for } (\hat{\lambda}_1, \hat{\lambda}_2) \in (R^C \cup L_1 \cup Z_1)^C \\ 0 & \text{otherwise.} \end{cases}$$

Finally, the estimator $\hat{\pi}_3$ follows:

$$\hat{\pi}_3^S = 1 - [\hat{\pi}_1^S + \hat{\pi}_2^S].$$

This shrinkage can be applied when the estimation of the proportion π_1 of the first group is more important and more sensible than the others. It is hence crucial to utilize in the estimation procedure all the information conveyed in the two samples about it. Figure 4 shows the values of the shrinkage estimator $\hat{\pi}_2^S$, as function of the values assumed by $\hat{\pi}_1$ and $\hat{\pi}_2$.

Remark 6. In the case where the most important proportion is related to the second group (group B), a different shrinkage procedure can be applied, giving more importance to the empirical result for the unknown proportion π_2 of group B. This new procedure can be performed, by the replacement of $\hat{\pi}_1$ with $\hat{\pi}_2$. Therefore, if needed, the first estimator to be shrunk is $\hat{\pi}_2$:

Fig. 4 The values of $\hat{\pi}_2^S$ for different values of the couple $(\hat{\pi}_1, \hat{\pi}_2)$

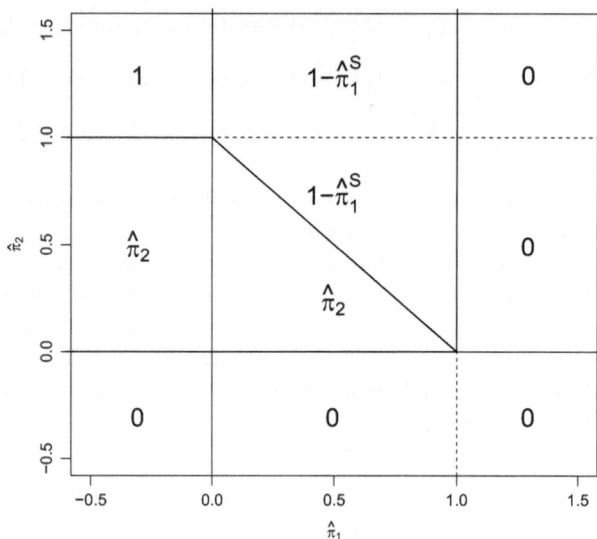

$$\hat{\pi}_2^S(\hat{\lambda}_1, \hat{\lambda}_2) = \begin{cases} 0 & \text{for}(\hat{\lambda}_1, \hat{\lambda}_2) \in L_2 \\ 1 & \text{for}(\hat{\lambda}_1, \hat{\lambda}_2) \in Z_2 \\ \hat{\pi}_2(\hat{\lambda}_1, \hat{\lambda}_2) & \text{otherwise.} \end{cases}$$

After that, just in case, the value of $\hat{\pi}_1$ changes as follows:

$$\hat{\pi}_1^S(\hat{\lambda}_1, \hat{\lambda}_2) = \begin{cases} 1 & \text{for}(\hat{\lambda}_1, \hat{\lambda}_2) \in L_2 \cap Z_1 \\ \hat{\pi}_1(\hat{\lambda}_1, \hat{\lambda}_2) & \text{for}(\hat{\lambda}_1, \hat{\lambda}_2) \in (R \cup L_1 \cup Z_1)^C \\ 1 - \hat{\pi}_2^S(\hat{\lambda}_1, \hat{\lambda}_2) & \text{for}(\hat{\lambda}_1, \hat{\lambda}_2) \in (R^C \cup L_2 \cup Z_2)^C \\ 0 & \text{otherwise.} \end{cases}$$

As before the value of $\hat{\pi}_3$ is then computed by:

$$\hat{\pi}_3^S = 1 - [\hat{\pi}_1^S + \hat{\pi}_2^S].$$

7 Conclusions

This paper describes a new procedure to obtain the estimators of the proportions of t population groups, which at least one is rare. As mentioned in the introduction, the description proposed in Sect. 2 deals with the trinomial case, but the extension to the case of more than three groups straightforwardly follows. The characteristics stated and proved in the previous sections highlight the potential of the estimators proposed and confirm that they can be successfully utilized in particular situations.

Indeed, this has to be considered a preliminary work, because further investigations are needed. A detailed comparison with other estimators and the comparison with other randomized response models are two examples of possible directions that must be explored in order to complete the analysis of these estimators, which look like very interesting and very useful in real situations.

References

Abul-Ela, A.L.A., Greenberg, B.G., Horvitz, D.G.: A multi-proportions randomized response model. J. Am. Stat. Assoc. **62**, 990–1008 (1967)

Best, D.J.: The variance of the inverse binomial estimator. Biometrika **61**, 385–386 (1974)

Cochran, W.G.: Sampling Techniques. Wiley, New York (1963)

Chaudhuri, A.: Randomized Response and Indirect Questioning Techniques in Surveys. CRC Press, Boca Raton (2011)

Chaudhuri, A., Bose, M., Dihidar, K.: Estimation of a sensitive proportion by Warners's randomized response data through inverse sampling. Stat. Pap. **52**, 343–354 (2011)

Chaudhuri, A., Bose, M., Dihidar, K.: Estimation sensitive proportions by Warners's randomized techinique using multiple randomized responses for distinct persons sampled. Stat. Pap. **52**, 111–124 (2011)

Finney, D.J.: On a method of estimating frequencies. Biometrika **36**, 233–234 (1949)

Gjestvang, C.R., Singh, S.: A new randomized response model. J. R. Stat. Soc. Ser. B **68**, 523–530 (2006)

Gupta, S., Gupta, B., Singh, S.: Estimation of sensitivity level of personal interview survey questions. J. Stat. Plan. Inference **100**, 239–247 (2002)

Haldane, J.B.S.: A labour-saving method of sampling. Nature (Lond.) **155**, 49–50 (1945)

Haldane, J.B.S.: On a method of estimating frequencies. Biometrika **33**, 222–225 (1945)

Kuk, A.Y.C.: Asking sensitive questions indirectly. Biometrika **77**, 436–438 (1990)

Mangat, N.S.: An improved randomized response strategy. J. R. Stat. Soc. Ser. B **56**, 93–95 (1994)

Mangat, N.S., Singh, R.: An alternative randomized response procedure. Biometrika **77**, 439–442 (1990)

Mangat, N.S., Singh, R.: Alternative approach to randomized response surveys. Statistica **51**, 327–332 (1991)

Mangat, N.S., Singh, R.: A note on the inverse binomial randomized response procedure. J. Indian Soc. Agric. Stat. **47**, 21–25 (1995)

Mikulski, P.W., Smith, P.J.: A variance bound for unbiased estimation in inverse sampling. Biometrika **63**, 216–217 (1976)

Pal, S., Singh, S.: A new unrelated question randomized response model. Statistics **46**, 99–109 (2012)

Pathak, P.K., Sathe, Y.S.: A new variance formula for unbiased estimation in inverse sampling. Sankhya Indian J. Stat. Ser. B **46**, 301–305 (1984)

Prasad, G., Sahai, S.: Sharper variance upper bound for unbiased estimation in inverse sampling. Biometrika **69**, 286 (1982)

Sahai, A.: Improved variance bounds for unbiased estimation in inverse sampling. J. Stat. Plan. Inference **4**, 213–216 (1980)

Sahai, A.: On a systematic sharpening of variance bounds of MVU estimator in inverse binomial sampling. Statistica **43**, 621–624 (1983)

Sathe, Y.S.: Sharper variance bounds for unbiased estimation in inverse sampling. Biometrika **64**, 425–426 (1977)

Singh, S.: A new stochastic randomized response model. Metrika **56**, 131–142 (2002)

Singh, H.P., Mathur, N.: An alternative randomized response technique using inverse sampling. Calcutta Stat. Assoc. Bull. **53**, 211–212 (2002)

Singh, H.P., Mathur, N.: On Mangat's improved randomized response strategy. Statistica **62**, 397–403 (2002)

Singh, H.P., Mathur, N.: On inverse binomial randomized response technique. J. Indian Soc. Agric. Stat. **59**, 192–198 (2005)

Singh, H.P., Mathur, N.: An improved estimation procedure for estimating the proportion of a population possessing sensitive attribute in unrelated question randomized response technique. Brasilian J. Probab. Stat. **20**, 93–110 (2006)

Sukhatme, P.V., Sukhatme, B.V., Sukhatme, S., Asok, C.: Sampling Theory of Surveys with Applications, 3rd edn. Iowa State University Press, Ames, Iowa (USA) and the Indian Society of Agricultural Statistics, New Delhi, India (1984)

Van der Heijden, P.G.M., Bockenholt, U.: Applications of randomized response methodology in e-commerce. In: Jank, W., Shmueli, G. (eds.) Statistical Methods in e-Commerce Research, pp. 401–416. Wiley, New York (2008)

Van der Heijden, P.G.M., Van Gils, G., Bouts, J., Hox, J.J.: A comparison of randomized response, computer-assisted self-interview, and face-to-face direct questioning. Sociol. Methods Res. **28**, 505–537 (2000)

Warner, S.L.: Randomized response: A survey technique for eliminating evasive answer bias. J. Am. Stat. Assoc. **60**, 63–69 (1965)

A Modified Extended Delete a Group Jackknife Variance Estimator Under Random Hot Deck Imputation in Business Surveys

Paolo Righi, Stefano Falorsi, and Andrea Fasulo

Abstract Item nonresponses commonly trouble large-scale surveys and if corrections are performed extra variability is introduced in the estimates. When the imputed values are treated as if they were observed the precision of the estimates is generally overstated. Modification of a variance estimator for contemplating item nonresponses is a ticklish issue. There is not a common judgement on which is the best estimation method. Usually the imputation procedure, the variance estimator properties and the cost-effectiveness issues lead to choose a specific method. The paper shows a method based on grouped jackknife easy to implement, not computer intensive and suitable with random hot deck imputation. A simulative comparison on real business data with the bootstrap method with imputed data and the Multiple Imputation has been carried out. In the simulation the sampling strategy of Italian Small and Medium Enterprises survey has been taken into account and Taylor linearization technique when imputed data are treated as true values has been considered as well. The findings show that the proposed method has good performances with respect to the other ones and it outperforms them in terms of computational time spending. This paper summarizes some results of the BLUE-Enterprise and Trade Statistics project (BLUE-ETS project—Work Package WP6, http://www.blue-ets.istat.it).

Keywords Bootstrap • Multiple Imputation • DAGJK • Monte Carlo simulation

P. Righi (✉) • S. Falorsi • A. Fasulo
Istat Italian National Statistical Institute, via Cesare Balbo 16, 00184 Roma, Italy
e-mail: parighi@istat.it; stfalors@istat.it; fasulo@istat.it

F. Mecatti et al. (eds.), *Contributions to Sampling Statistics*, Contributions to Statistics,
DOI 10.1007/978-3-319-05320-2__14,
© Springer International Publishing Switzerland 2014

1 Introduction

Variance estimation has to take into account additional complexity elements: the unit and item nonresponse that commonly trouble the large-scale surveys. Unit nonresponse is customarily handled by forming weighting classes using auxiliary variables observed on all the sampled elements and then adjusting the survey weights of all respondents within a weighting class by a common nonresponse adjustment factor, with different adjustment factors in different classes (Kalton and Kasprzyk 1986).

Imputation is the commonly used approach to compensate for missing (item nonresponse) or invalid values in sample surveys (Kalton and Kasprzyk 1986).

When unit and item nonresponse correction is performed extra variability is introduced in the sampling errors. Modifications of Taylor and resampling methods for contemplating unit nonresponse are quite straightforward while item nonresponse is a ticklish issue. Analyses performed on imputed values treated as if they were observed can be misleading when estimates of the variance do not include the variability component due to imputation. As a result, the precision of estimates is overstated, and subsequent statistical analyses can be misleading (e.g., confidence intervals have lower than nominal levels).

The approaches proposed in literature to obtain valid variance estimators in presence of imputed data are divided according to several classifications. A first common classification distinguishes among linearization methods or Model Assisted techniques (Sarndal 1992), the resampling methods (Shao and Tu 1995; Wolter 2007) and the Multiple Imputation method (Rubin 1987; Little and Rubin 2002), being the first two categories used for the complete data variance estimators as well.

The resampling techniques in presence of item nonresponse can by subdivided according to the standard classification used for the complete data (Wolter 2007). Then, bootstrap (Efron 1994; Shao and Sitter 1996; Saigo et al. 2001; Shao 2003), balanced repeated replication (Rao and Shao 1999), random group methods (Shao and Tang 2011) and jackknife (Rao and Shao 1992; Rao 1996; Yung and Rao 2000; Chen and Shao 2001; Skinner and Rao 2002; Saigo and Sitter 2005) can be distinguished.

One of the main drawbacks with the replication methods is the computational effort because of the number of replications. To deal with this problem we propose a grouped jackknife taking into account the random hot deck imputation. The grouped jackknife methods delete groups of units rather than one unit at time for reducing calculation effort. The creation of a replicate group can be done within design stratum or, combining design strata into superstrata, taking groups of units within a superstratum that cut across design strata (Rust 1985, 1986; Rust and Rao 1996).

The only guidance is that the sample should be divided into specially design replicate subsamples that mirror the design of the full sample. A common assumption in the literature is to form equal sized groups. Moreover, to take into account a nonnegligible sampling fraction, it can be useful to form superstrata with design strata having similar sampling fractions (Valliant et al. 2008). Finally, grouped jackknife methods distinguish themselves by the computation of the replicate weights as well, augmenting the possible variance estimators.

Currently there is no empirical evidence showed in literature, suggesting the best grouped jackknife. Then we should underline the importance of the practical application of such methods in the Official Statistics context.

As concerns grouped jackknife methods taking into account item nonresponse few have been written in the literature. Brick et al. (2005) propose a grouped adjusted jackknife according to Rao and Shao approach in case of Horvitz–Thompson estimator. Di Zio et al. (2008) give some suggestions to modify the Delete a Group Jackknife (DAGJK) proposed by Kott (1998, 2006) for dealing with the case of imputed values. Miller and Kott (2011) investigate a DAGJK with imputed data, too.

In Sect. 2 we show a new method based on the Extended Delete a Group Jackknife (EDAGJK) proposed by Kott (2001) involving the adjustment by Rao and Shao (1992). The proposed adjustment for imputation holds also for DAGJK method. In Sect. 3 the modified EDAGJK for imputation is compared in a simulation study on real survey data with the MI and bootstrap under imputation. The conclusions of empirical evaluations (Sect. 4) state that the modified EDAGJK produces nearly unbiased variance estimates and it works well with respect to the two benchmarking methods such as MI and bootstrap under imputation.

2 The Modified Extended DAGJK Under Rao and Shao Adjustment for Imputation

Delete A Group Jackknife (Kott 1998, 2001) is a variance estimation technique computationally less intensive than classical jackknife, then it can be applied also in case of large-scale surveys. DAGJK is within the strategies aiming at reducing the number of jackknife replications, while maintaining adequate precision of variance estimates. It assumes an unique superstratum formed by all the design strata and the replicate groups have units belonging to different design strata. Then, the method does not present implications on the definition of the groups and does not require analysis to form superstrata. This analysis can become cumbersome for large-scale and complex business surveys and may affect the timeliness of data production. For this reason variance estimation techniques implemented by a sort of automated process and leading to *good* statistical results may be preferred to better techniques but more complex to be implemented.

Let's consider the stratified simple random sampling commonly used in the business surveys, where in each stratum h, are included N_h units. A sample of $n_h \geq 2$ units is drawn from each stratum independently across strata. Let $d_{hk}(> 0)$

be the basic weight of unit k in stratum h ($h = 1, \ldots, L$), denoted as hk, the estimator of the parameter of the total θ is $\hat{\theta} = \sum_{hk \in s} d_{hk} y_{hk}$, being y_{hk} the value of the variable of interest for unit i. The DAGJK technique divides the overall or *parent* sample s into Q mutually exclusive *replicate* or *random* groups, hereinafter denoted by $s^1, \ldots, s^q, \ldots, s^Q$. Given the subsample s^q, the sample sizes in the strata are indicated as $n_1^q, \ldots, n_h^q, \ldots, n_L^q$. The complement of each s^q is called the *jackknife replicate group* $s^{(q)} = s - s^q$, being $n_1^{(q)}, \ldots, n_h^{(q)}, \ldots, n_L^{(q)}$ the strata sample sizes of $s^{(q)}$. The variance estimator is based on the following jackknife steps:

1. Units are randomly ordered in each stratum.
2. From this ordering the units are systematically allocated into Q groups.
3. For each unit hk, Q different sampling weights (*replicate sampling weights*) are computed.
4. Given the qth set of replicate weights the qth replicate estimate is
   $\hat{\theta}^{(q)} = \sum_{hk \in s} d_{hk}^{(q)} y_{hk}$ where $d_{hk}^{(q)}$ denotes the qth replicate weight of unit hkth.
5. The DAGJK variance estimator is given by

$$v(\hat{\theta}) = \frac{Q-1}{Q} \sum_{q=1}^{Q} (\hat{\theta}^{(q)} - \hat{\theta})^2. \tag{1}$$

Kott developed different expression of the replicate weights. In particular, when $n_h < Q$, to perform an unbiased variance estimation Kott proposed the EDAGJK (Kott 2001) when the finite population correction is negligible. For the Horvitz–Thompson estimator the replicate weights of EDAGJK assume the following expression:

$$d_{hk}^{(q)} = \begin{cases} d_{hk}, & \text{when } k \in h \text{ and no units of } h \text{ belongs to group } q; \\ d_{hk}[1 - (n_h - 1)Z], & \text{when } k \in q; \\ d_{hk}(1 + Z), & \text{otherwise}, \end{cases} \tag{2}$$

where $Z^2 = Q/[(Q-1)n_h(n_{h-1})]$.

With the Greg estimator the replicate weights are given by $w_{hk}^{(q)} = d_{hk}^{(q)} \gamma_{hk}^{(q)}$ and the replicate qth GREG estimate is $\hat{\theta}_{greg}^{(q)} = \sum_{hk \in s} y_{hk} w_{hk}^{(q)}$. The correction factor $\gamma_{hk}^{(q)}$ may be calculated according to different ways. Let us consider the following expressions:

$$\gamma_{hk}^{(q)} = 1 + \left(\mathbf{X} - \sum_{hk \in s} \mathbf{x}_{hk} d_{hk}^{(q)} \right) \left(\sum_{hk \in s} \frac{\mathbf{x}_{hk} \mathbf{x}_{hk}' d_{hk}^{(q)}}{c_{hk}} \right)^{-1} \frac{\mathbf{x}_{hk}}{c_{hk}} \tag{3}$$

and

$$\gamma_{hk}^{(q)} = \gamma_{hk} + \left(\mathbf{X} - \sum_{hk \in s} \mathbf{x}_{hk} d_{hk}^{(q)} \gamma_{hk} \right) \left(\sum_{hk \in s} \frac{\mathbf{x}_{hk} \mathbf{x}_{hk}' d_{hk}^{(q)} \gamma_{hk}}{c_{hk}} \right)^{-1} \frac{\mathbf{x}_{hk} \gamma_{hk}}{c_{hk}}. \qquad (4)$$

Kott (1998, 2006) offers some suggestions on the factor has to be used.

In order to take into account item nonresponse in variance estimation, we propose a modified version of EDAGJK based on the Rao and Shao adjustment for hot deck imputation.

The modified variance estimator is

$$v(\hat{\theta}_I) = \frac{Q-1}{Q} \sum_{q=1}^{Q} (\hat{\theta}_I^{(q)} - \hat{\theta}_I)^2, \qquad (5)$$

where

$$\hat{\theta}_I = \sum_{hk \in s_R} w_{hk} y_{hk} + \sum_{hk \in s_{\bar{R}}} w_{hk} y_{hk}^* \qquad (6)$$

is the estimator with imputed hot deck values y_{hk}^*, being s_R and $s_{\bar{R}}$ the sample of respondents and non respondents.

$\hat{\theta}_I^{(q)}$ is defined as

$$\hat{\theta}_I^{(q)} = \sum_{g=1}^{G} \left\{ \sum_{hk \in s_{R_g}} w_{hk}^{(q)} y_{hk} + \sum_{hj \in s_{\bar{R}_g}} w_{hj}^{(q)} \left(y_{hj}^* + \hat{\bar{y}}_{R_g}^{(q)} - \bar{y}_{R_g} \right) \right\}, \qquad (7)$$

in which: g $(g = 1, \ldots, G)$ indicates the gth imputation cell; s_{R_g} and $s_{\bar{R}_g}$ are, respectively, the respondents and non-respondents in the cell g; $w_{hk}^{(q)}$ are the replicate base or Greg weights. Finally, $\hat{\bar{y}}_{R_g}^{(q)} = \sum_{hj \in s_{R_g}} w_{hj}^{(q)} y_{hj} / \sum_{hj \in s_{R_g}} w_{hj}^{(q)}$ and $\bar{y}_{R_g}^{(q)} = \sum_{hj \in s_{R_g}} w_{hj} y_{hj} / \sum_{hj \in s_{R_g}} w_{hj}$.

Note that the imputation procedure is performed only on the parent sample.

3 Empirical Study

In the literature there is not a common judgement on which is the best approach or method for variance estimation with incomplete data. Usually the best method depends on the survey context involving the sampling design, the estimator, the type and the number of parameters to be estimated, the domains of interest, the

Table 1 Summary of the target variables

Variables	Min.	1st Qu.	Median	Mean	3rd Qu.	Max.
CIF	−869,300	−3,750	−130	−2,422	1,490	322,700
PUC	−6,631	5,514	19,710	41,880	56,230	247,900
OEX	0	441	1,601	6,201	5,294	126,200
LCO	0	0	0	16,700	24,270	126,500

imputation process. In general the imputation procedure, the estimated function and cost-effectiveness issues lead to choose a specific method. In this section we compare the performances of the modified EDAGJK with bootstrap and MI in the context of a typical sampling strategy of a business survey.

3.1 The Population and the Sampling Strategy

The population in the simulation study based on the real data of the 2008 Italian enterprises belonging to the economic activity 162 according to the Statistical Classification of Economic Activities in the European Community NACE Rev.2 3-digit (number of units $N = 21,231$). This subpopulation is surveyed in the small and medium enterprises (SME) survey. The SME is a yearly survey investigating the profit-and-loss account of enterprises with less than 100 employed persons, as requested by SBS EU Council Regulation n. 58/97 (Eurostat, 2003) and n. 295/2008. The Italian target population of the SME survey is about 4.5 millions active enterprises.

The following target variables have been considered: Changes of inventory of finished and semifinished products (CIF); Purchase of commodities (PUC); Operating expenses for administration (OEX); Labour cost (LCO). The values of the four variables have been taken from the balance sheets (administrative data) for the whole population.

Table 1 gives some summary statistics.

The simulation takes into account of a simplified version of the current sampling strategy used in the SME survey. A stratified simple random sampling design and a calibration estimator have been considered.

Strata are obtained by crossing the size classes and the regions according to the Nomenclature of territorial units for statistics NUTS 1 defined by EU. Hence, 20 strata are obtained as aggregation of the original strata of SME survey. The sample allocation in each stratum is taken from the allocation of 2008 SME survey. Table 2 shows the population and sample distribution in each stratum. The overall sample size is $n = 908$ enterprises.

The estimator calibrates the sampling weights to the number of enterprises and the number of employed persons at NACE Rev.2 4-digit, size class and NUTS 1 region.

Table 2 Population, sample distribution and missing rates (%) in the design strata

Strata (NUTS 1 region by size class)	Number of enterprises in the population	Number of enterprises allocated in the sample	Sampling rate	Missing rate by variable			
				CIF	PUC	OEX	LCO
NORTH-WEST:(size class<8)	5,241	107	0.02	16.75	1.28	7.52	1.97
NORTH-WEST:(9<size class<18)	425	39	0.09	14.59	1.41	3.29	0.00
NORTH-WEST:(19<size class<28)	83	17	0.20	28.92	2.41	8.43	0.00
NORTH-WEST:(size class>29)	38	15	0.39	13.16	5.26	5.26	0.00
NORTH-EAST:(size class<8)	5,087	89	0.02	18.28	2.54	9.44	3.07
NORTH-EAST:(9<size class<18)	550	48	0.09	14.18	3.45	6.00	0.18
NORTH-EAST:(19<size class<28)	129	44	0.34	15.50	4.65	7.75	0.00
NORTH-EAST:(size class>29)	55	20	0.36	19.35	3.23	3.23	0.00
CENTER:(size class<8)	3,699	145	0.04	19.60	1.14	9.22	1.57
CENTER:(9<size class<18)	297	34	0.11	19.53	2.69	5.05	0.00
CENTER:(19<size class<28)	63	35	0.56	19.35	3.23	3.23	0.00
CENTER:(size class>29)	31	17	0.55	20.79	2.60	10.43	2.78
SOUTH:(size class<8)	3,386	128	0.04	17.61	1.14	5.11	0.00
SOUTH:(9<size class<18)	176	45	0.26	13.51	0.00	0.00	0.00
SOUTH:(19<size class<28)	37	23	0.62	26.67	0.00	0.00	0.00
SOUTH:(size class>29)	15	8	0.53	18.64	1.16	10.21	2.83
ISLANDS:(size class<8)	1,803	50	0.03	13.48	0.00	3.37	1.12
ISLANDS:(9<size class<18)	89	26	0.29	4.76	0.00	0.00	0.00
ISLANDS:(19<size class<28)	21	16	0.76	4.76	0.00	0.00	0.00
ISLANDS:(size class>29)	6	2	0.33	16.67	16.67	16.67	16.67
		Missing rate	Average	18.36	1.88	8.73	2.19

The linear distance function (generalized regression estimator) is considered. Actually, the logit distance function is used in the SME survey, because it produces nonnegative calibrated weights. Nevertheless the logit distance has two drawbacks: the convergence is not guaranteed; it requires a time spending iterative procedure to obtain the calibrated weights. For these reasons linear distance function has been preferred in the simulation. We point out that the calibration estimators with linear and logit distance function converge asymptotically and the simulation results with the linear distance will be coherent with the ones obtained with the logit distance function.

The main task of the simulation study is to compare different methods of variance estimation for estimators of totals in a complex context, usual in the business survey, such as: stratified simple random sampling, imputation for item nonresponse and calibration estimator. The item nonresponses are imputed by means of random hot-deck procedure.

Besides this sampling strategy the simulation regards the variance of the Horvitz–Thompson (HT) estimator.

3.2   Item Nonresponse Model

The item nonresponses have been generated seeking to reproduce the item nonresponse pattern of the business surveys.

The SME survey suffers from item nonresponses. Actually the survey has not flags for item nonresponses and the item nonresponses are denoted by zero values.

To find out when a zero value means a real zero or a missing value, the 2008 SME data have been linked with the 2008 administrative data; the zero SME values corresponding to a non-zero value in the administrative data have been identified as missing values.

A regression tree model (rpart R package) has been applied for estimating the relationship between response propensity and outcome-related auxiliary variables known for the whole population. To create missing values, response indicators were assigned to the units within nonresponse cells defined by the regression tree. Within a given response cell, units were assigned at random to be missing or nonmissing at a specified rate. The mechanism generating missingness assumes that there is a uniform response probability within each cell. This is an usual assumption for the nonresponse model even though generally a more complex unknown item nonresponse model holds. When the real nonresponse model disagrees with the working model used for the imputation, the estimates are biased. However, the simulation is focused on the variance estimate and then we define an experimental context in which the point estimates are unbiased, for not creating confounding evidences.

Table 2 shows the strata missing rates (the last four columns) for the variables of interest. Then, three type rates of item nonresponse appear: high for the CIF variable

Table 3 Relative bias(RB$(\widehat{\theta}_I)\%$) of the estimators

Estimators	CIF	PUC	OEX	LCO
HT with imputation	0.72	0.49	0.94	−0.09
CALIBRATION with imputation	1.40	0.51	0.97	−0.10

with average equal to 18.36 %, medium for the OEX variable with 8.73 % and low for the PUC and LCO variables with about 2 % of nonresponse rate.

We checked the unbiasedness of the estimator after the hot deck imputation computing the empirical relative bias

$$\mathrm{RB}(\widehat{\theta}_I) = \frac{1}{C} \sum_{c=1}^{C} \frac{(\widehat{\theta}_{Ic} - \theta)}{\theta}, \tag{8}$$

being $\widehat{\theta}_{Ic}$ the estimate from the sample c drawn according to the sampling design of Sect. 3.1 and C the number of drawn samples. To obtain a nearly zero relative bias $C = 10{,}000$ samples have been selected.

Table 3 shows negligible bias for all estimates: e.g. the RB$(\widehat{\theta}_I)\%$ are lower than 1 % except for the variable CIF when calibration with imputed data is considered (1.4 %).

3.3 Results of the Monte Carlo Simulation

Several methods are compared in the simulation. Furthermore, the following reference variances

$$V(\widehat{\theta}_I) = \sum_{c=1}^{10{,}000} \frac{(\widehat{\theta}_{Ic} - \theta)^2}{10{,}000}, \tag{9}$$

hereinafter denoted as empirical or Monte Carlo variances, are computed.

For the HT estimator the following are considered:

- Unbiased variance estimator (Wolter 2007) denoted as STANDARD method.
- EDAGJK according to Kott (2001).
- EDAGJK.I: the modified EDAGJK (Sect. 2) using the replicate weights given in the (2).
- BOOTSTRAP.I: bootstrap variance methods under imputation (Shao and Sitter 1996).
- MI: using the Approximate Bayesian Bootstrap (ABB) (Kim and Fuller 2004; Brick et al. 2005).

For the calibration estimator the following have been compared:

- TAYLOR variance estimator.
- EDAGJK.HT computed according to (3).
- EDAGJK.CAL computed according to (4).
- EDAGJK.HT.I: the modified EDAGJK (Sect. 2) based on the correction factor (3).
- EDAGJK.CAL.I: the modified EDAGJK (Sect. 2) based on the correction factor (4).
- BOOTSTRAP.I: bootstrap variance methods under imputation (Shao and Sitter 1996).
- MI: using the ABB.

Note that the STANDARD, EDAGJK, TAYLOR, EDAGJK.HT and EDAGJK.CAL, do not properly take into account the imputation correction.

The complete simulation has implemented twelve variance methods by four variables. The imputation procedure is performed and for seven variance estimators imputation adjustment is carried out as well. Then, computational issues led us to choose 1,000 replications in performing the variance estimates for each method.

The accuracy of the variance estimates is measured with the following summary statistics:

- The Relative (percentage) Bias of Variance estimation

$$
\mathrm{RB}[v(\widehat{\theta}_I)]\% = 100 \times \frac{\bar{v}(\widehat{\theta}_I) - V(\widehat{\theta}_I)}{V(\widehat{\theta}_I)}. \tag{10}
$$

- The Relative (percentage) Root Mean Square Error of Variance estimation

$$
\mathrm{RRMSE}[v(\widehat{\theta}_I)]\% = 100 \times \sqrt{\frac{\frac{1}{1,000} \sum_{c=1}^{1,000} \left[v(\widehat{\theta}_{Ic}) - V(\widehat{\theta}_I) \right]^2}{V(\widehat{\theta}_I)^2}}. \tag{11}
$$

- The Coverage of the Confidence Interval (percentage), that is the percentage of intervals including θ, based on the nominal 95 % confidence intervals computed for each of 1,000 simulations. We used the normal distribution as approximation of the t distribution

$$
\mathrm{CCI}[v(\widehat{\theta}_I)]\% = \frac{100}{1,000} \sum_{c=1}^{1,000} \delta_c \text{ where } \delta_c = 1 \text{ if } \theta \in \left(\widehat{\theta}_{Ic} \pm 1.96 \sqrt{v(\widehat{\theta}_{Ic})} \right)
$$

and $\delta_c = 0$ otherwise.

- The Lower Error Rate and Upper Error Rate

$$
\mathrm{LER}[v(\widehat{\theta}_I)]\% = 100 \times \frac{1}{1,000} (\text{number of samples with} \quad \theta < -1.96\sqrt{v(\widehat{\theta}_{Ic})}),
$$

$$
\mathrm{UER}[v(\widehat{\theta}_I)]\% = 100 \times \frac{1}{1,000} (\text{number of samples with} \quad \theta > +1.96\sqrt{v(\widehat{\theta}_{Ic})}).
$$

Table 4 Relative Bias $(\mathrm{RB}[v(\widehat{\theta}_I)])$ and Relative Root Means Square Error $(\mathrm{RRMSE}[v(\widehat{\theta}_I)])$ of the variance estimators with imputed data

Variance estimator	H-T estimator							
	$\mathrm{RB}[v(\widehat{\theta}_I)]\%$				$\mathrm{RRMSE}[v(\widehat{\theta}_I)]\%$			
	CIF	PUC	OEX	LCO	CIF	PUC	OEX	LCO
STANDARD	−27.15	−3.37	−8.96	−6.40	53.43	38.54	204.64	15.41
EDAGJK	−21.26	3.11	−8.11	−0.30	57.94	45.90	178.76	31.26
EDAGJK.I	7.43	7.44	6.77	3.99	71.35	47.65	209.31	32.61
BOOTSTRAP.I	7.95	5.57	21.40	2.81	64.53	42.99	261.52	21.90
MI	−2.83	2.36	−1.82	−3.08	66.74	40.96	205.00	15.58
	Calibration estimator							
	$\mathrm{RB}[v(\widehat{\theta}_I)]\%$				$\mathrm{RRMSE}[v(\widehat{\theta}_I)]\%$			
TAYLOR	−28.48	−6.21	−9.82	−7.55	53.90	43.93	227.47	23.82
EDAGJK.HT	−19.76	3.60	−5.73	2.95	59.27	51.12	195.25	36.73
EDAGJK.CAL	−18.24	6.79	−6.56	4.98	62.15	55.04	186.05	40.89
EDAGJK.HT.I	9.48	8.01	9.15	6.61	73.30	53.02	225.78	38.60
EDAGJK.CAL.I	11.51	11.31	8.23	8.75	77.73	56.85	215.29	42.76
BOOTSTRAP.I	5.71	6.57	14.43	4.79	64.38	48.84	244.40	27.39
MI	−9.05	−4.46	−15.34	0.17	57.24	42.46	177.88	22.42

Table 4 shows that for the variables with large nonresponse rate (CIF and OEX), the methods that do not take properly into account the imputation process such as STANDARD, EDAGJK, TAYLOR, EDAGJK.HT and EDAGJK.CAL produce large downward biased variance estimates. The result was definitely expected.

Furthermore for the OEX variable we observe a very large variability with the $\mathrm{RRMSE}[v(\widehat{\theta}_I)]\%$ over than 177 % for all the methods. The evidence is explained by the positive skew distribution of this variable (Fig. 1).

The scatterplot of the 10,000 variance estimates versus the corresponding HT estimates (Fig. 2) shows for the OEX variable two separate clouds, being the highest one around the size of 200. This is due to one extreme value within the stratum NORTH-WEST:(size class<8) with the 0.02 sampling rate. For this stratum the expected percentage contribution to the overall variance is around 85 %. Furthermore, the stratum variance when the extreme value is included is about five times the stratum variance when the extreme value is not included in the sample.

The presence of rare extreme values is typical in the business surveys and then it is interesting to study the behaviour of the estimators in this critical context.

In the following the main comments are focused on CIF variable, because it has the highest missing rate. As concerns the HT estimator EDAGJK.I and bootstrap methods, they produce $\mathrm{RB}[v(\widehat{\theta}_I)]\%$ around 7 % but bootstrap has a smaller $\mathrm{RRMSE}[v(\widehat{\theta}_I)]\%$ than EDAGJK.I. MI has the smallest $\mathrm{RB}[v(\widehat{\theta}_I)]\%$ with the drawback to be negative.

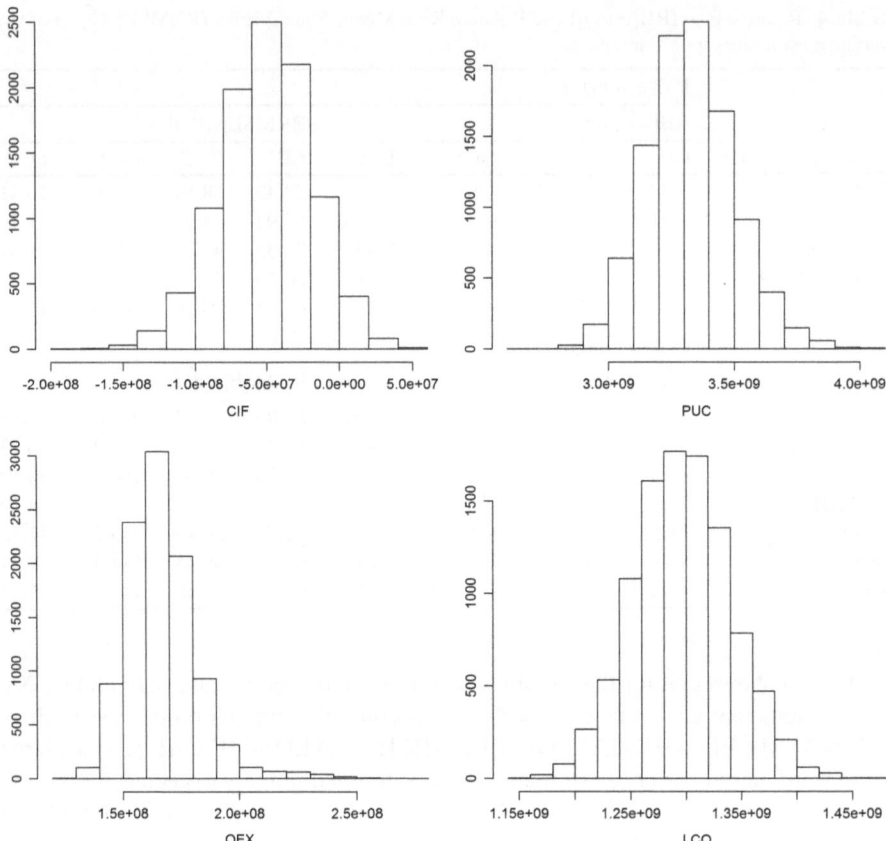

Fig. 1 Distribution of the 10,000 HT estimates

In case of calibration estimator, bootstrap outperforms the methods that consider specifically the imputation in terms of $RB[v(\widehat{\theta_I})]\%$. Nevertheless the modified EDAGJK methods produce positive and not so large $RB[v(\widehat{\theta_I})]\%$. MI has negative bias, but outperforms the other estimators in terms of $RRMSE[v(\widehat{\theta_I})]\%$. Table 4 shows that EDAGJK.HT is slightly better than EDAGJK.CAL at least for the CIF variable.

Table 5 shows the coverage of the confidence interval. The methods ignoring that many values are imputed have a strong reduction of the coverage rates. MI does not show good performances, at least for the CIF variable, while a small decreasing of coverage is observed for the rest of the resampling methods. The modified EDAGJK techniques and bootstrap are essentially equivalent. For the calibration estimator, EDAGJK.CAL.I seems slightly better than EDAGJK.HT.I and bootstrap. That occurs because of a larger $RB[v(\widehat{\theta_I})]\%$.

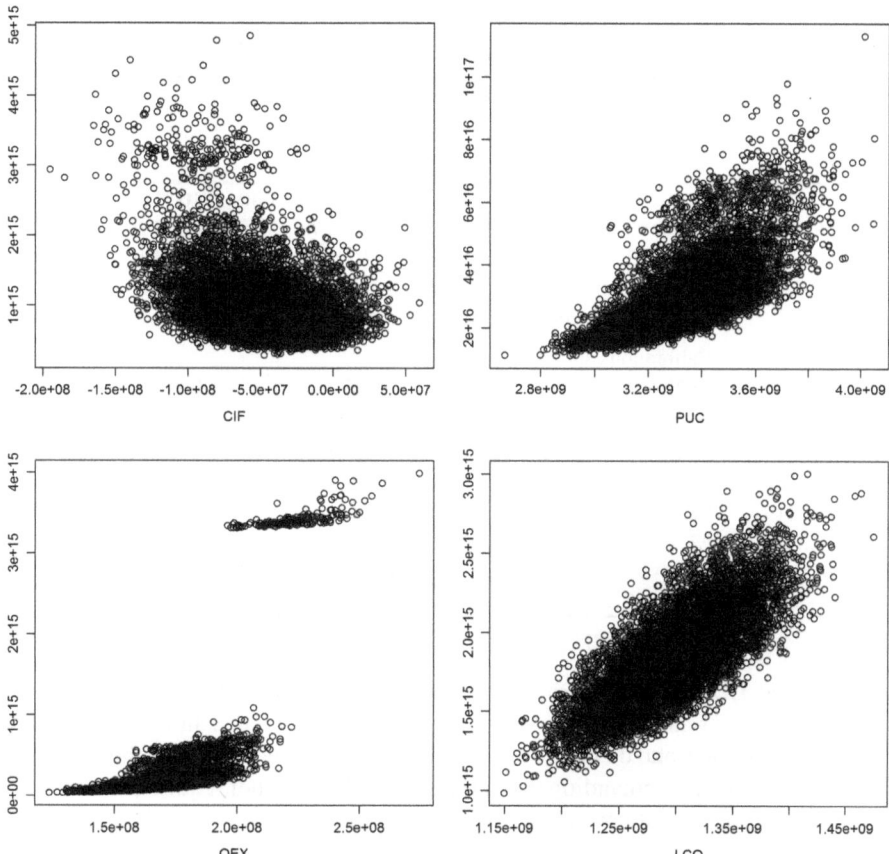

Fig. 2 Scatterplot of the HT estimates versus standard variance estimates

4 Conclusions

The paper shows a new variance estimator taking into account the imputation process. It combines the EDAGJK technique proposed by Kott with the adjusted jackknife proposed by Rao and Shao. The reasons leading to this choice is the good compromise among theoretical properties and practical aspects. In particular EDAGJK produces an unbiased estimator (for complete data set), it is easy to implement and not computer intensive. Furthermore, the adjusted jackknife does not require replications of the imputation procedure.

These features are quite appealing especially in a National Statistical Institute, where data production based on large data sets must be automatized as much as possible.

The method has been compared via simulation based on real business data with two standard techniques: MI and bootstrap. The simulation results show that

Table 5 Coverage of the confidence interval(CCI[$v(\widehat{\theta}_I)$]), the Lower and Upper Error Rate (LER[$v(\widehat{\theta}_I)$], UER[$v(\widehat{\theta}_I)$]) with imputed data

	H-T estimator											
	CCI[$v(\widehat{\theta}_I)$]%				LER[$v(\widehat{\theta}_I)$]%				UER[$v(\widehat{\theta}_I)$]%			
Variance estimator	CIF	PUC	OEX	LCO	CIF	PUC	OEX	LCO	CIF	PUC	OEX	LCO
STANDARD	88.40	94.50	90.40	94.40	7.50	1.20	1.00	1.40	4.10	4.30	8.60	4.20
EDAGJK	86.20	93.00	88.20	94.00	6.80	4.70	7.70	3.40	7.00	2.30	4.10	2.60
EDAGJK.I	91.10	93.70	89.90	94.40	4.20	4.20	6.90	3.10	4.70	2.10	3.20	2.50
BOOTSTRAP.I	92.30	93.80	90.30	95.90	2.90	4.20	7.10	2.30	4.80	2.00	2.60	1.80
MI	89.90	94.40	90.60	95.50	4.50	3.90	7.20	2.50	5.60	1.70	2.20	2.00
	Calibration estimator											
	CCI[$v(\widehat{\theta}_I)$]%				LER[$v(\widehat{\theta}_I)$]%				UER[$v(\widehat{\theta}_I)$]%			
TAYLOR	88.20	94.20	90.20	93.40	7.70	1.80	0.90	2.00	4.10	4.00	8.90	4.60
EDAGJK.HT	86.20	92.70	89.40	93.60	6.50	5.30	7.00	3.00	7.30	2.00	3.60	3.40
EDAGJK.CAL	86.76	93.88	89.87	92.98	6.32	4.61	7.12	3.51	6.92	1.50	3.01	3.51
EDAGJK.HT.I	91.80	92.90	91.20	93.70	3.50	5.20	6.30	3.00	4.70	1.90	2.50	3.30
EDAGJK.CAL.I	92.38	94.18	91.57	93.68	3.71	4.41	5.82	3.21	3.91	1.40	2.61	3.11
BOOTSTRAP.I	92.00	93.72	90.08	95.65	3.24	4.66	6.88	1.92	4.76	1.62	3.04	2.43
MI	90.50	92.20	88.20	93.50	4.00	6.00	8.40	3.30	5.50	1.80	3.40	3.20

the modified EDAGJK with Rao and Shao adjustment produces nearly unbiased variance estimates and it works well with respect to the two benchmarking methods in terms of accuracy and coverage of confidence interval. Nevertheless the method is less computationally demanding than bootstrap and it does not require an increasing of complexity of the data production process as for MI.

References

Brick, J.M., Jones, M.E., Kalton, G., Valliant, R.: Variance estimation with Hot Deck imputation: a simulation study of three methods. Surv. Methodol. **31**, 151–159 (2005)

Chen, J., Shao, J.: Jackknife variance estimation for nearest-neighbour imputation. J. Am. Stat. Assoc. **95**, 260–269 (2001)

Di Zio, M., Falorsi, S., Guarnera, U., Luzi, O., Righi, P.: Variance estimation in presence of imputation: an application to an Istat survey data. In: Proceedings of the Section on Survey Research Methods European Conference on Quality in Official Statistics (2008). http://q2008. istat.it/sessions/paper/17DiZio.pdf

Efron, B.: Missing data, imputation and the bootstrap. J. Am. Stat. Assoc. **89**, 463–479 (1994)

Kalton, G., Kasprzyk, D.: The treatment of missing survey data. Surv. Methodol. **12**, 1–16 (1986)

Kim, J.K., Fuller, W.A.: Fractional Hot Deck imputation. Biometrika **91**, 559–578 (2004)

Kott, P.: Using the delete-a-group jackknife variance estimator in NASS surveys. Nass Research Report 98–01 (revised 2001), NASS (1998). http://www.nass.usda.gov/research/ reports/RRGJ7.pdf

Kott, P.: Delete-a-group jackknife. J. Off. Stat. **17**, 521–526 (2001)

Kott, P.: Delete-a-group variance estimation for the general regression estimator under poisson sampling. J. Off. Stat. **22**, 759–767 (2006)

Little, R.J.A., Rubin, D.B.: Statistical Analysis with Missing Data. Wiley Series in Probability and Statistics. Wiley, London (2002)

Miller, D., Kott, P.: Using the DAG jackknife to measure the variance of an estimator in the presence of item nonresponse. In: JSM Proceedings, Statistical Computing Section. American Statistical Association, Alexandria (2011)

Rao, J.N.K.: On variance estimation with imputed survey data. J. Am. Stat. Assoc. **91**, 499–506 (1996)

Rao, J.N.K., Shao, S.: Jackknife variance estimation with survey data under Hot Deck imputation. Biometrika **79**, 811–822 (1992)

Rao, J.N.K., Shao, S.: Modified balanced repeated replication for complex survey data. Biometrika **86**, 403–415 (1999)

Rubin, D.B.: Multiple Imputation for Nonresponse in Surveys. Wiley Series in Probability and Statistics. Wiley, London (1987)

Rust, K.: Variance estimation for complex estimators in sample surveys. J. Off. Stat. **1**, 381–397 (1985)

Rust, K.: Efficient formation of replicates for replicated variance estimation. In: Proceedings of the Section on Survey Research Methods of the American Statistical Association, pp. 81–87 (1986)

Rust, K., Rao, J.N.K.: Variance estimation for complex surveys using replication techniques. Stat. Methods Med. Res. **5**, 283–310 (1996)

Saigo, H., Shao, J., Sitter, R.R.: A repeated half-sample bootstrap and balanced repeated replications for randomly imputed data. Surv. Methodol. **27**, 189–196 (2001)

Saigo, H., Sitter, R.R.: Jackknife variance estimator with reimputation for randomly imputed survey data. Stat. Probab. Lett. **73**, 321–331 (2005)

Shao, J.: Impact of the bootstrap on sample surveys. Stat. Sci. **18**, 191–198 (2003)

Shao, J., Sitter, R.R.: Bootstrap for imputed survey data. J. Am. Stat. Assoc. **91**, 1278–1288 (1996)

Shao, J., Tang, Q.: Random group variance estimators for survey data with random Hot Deck imputation. J. Off. Stat. **27**, 507–526 (2011)

Shao, J., Tu, D.: The Jacknife and Bootstrap. Springer, Berlin (1995)

Skinner, C.J., Rao, J.N.K.: Jackknife variance estimation for multivariate statistics under Hot-Deck imputation from common donor. J. Stat. Plan. Inference **102**, 149–167 (2002)

Sarndal, C.E.: Methods for estimating the precision of survey estimates when imputation has been used. Surv. Methodol. **18**, 241–252 (1992)

Valliant, R., Brick, M.J., Dever, J.: Weight adjustments for the grouped jackknife variance estimator. J. Off. Stat. **24**, 469–488 (2008)

Wolter, K.: Introduction to Variance Estimation. Springer, London (2007)

Yung, W., Rao, J.N.K.: Jackknife variance estimation under imputation for estimators using poststratification information. J. Am. Stat. Assoc. **95**, 903–915 (2000)